Community School Mathematics 2B

Pat Lilburn
Pam Rawson
Peter Sullivan

Oxford University Press
Department of Education, Papua New Guinea

OXFORD UNIVERSITY PRESS AUSTRALIA

Oxford New York
Athens Auckland Bangkok Bombay
Calcutta Cape Town Dar es Salaam Delhi
Florence Hong Kong Istanbul Karachi
Kuala Lumpur Madras Madrid Melbourne
Mexico City Nairobi Paris Singapore
Taipei Tokyo Toronto

and associated companies in
Berlin Ibadan

OXFORD is a trade mark of Oxford University Press

First published 1992
Reprinted 1993, 1994, 1997, 1999, 2000, 2014(D)

ISBN 978 9980 58854 8

Written by Pat Lilburn, Pam Rawson and Peter Sullivan
Writing consultant Elsie Kinavai
Papua New Guinea project co-ordinator and writing consultant Katherine Schneider

Cover photograph by Rocky Roe Photographics
Illustrated by Juli Kent, Annie Vanston and Mary Ann Hurley
Typeset by Solo Typesetting, South Australia
Printed and bound in Australia by Ligare Book Printers, Pty Ltd
Published by Oxford University Press,
253 Normanby Road, South Melbourne, Australia and
The Department of Education, Papua New Guinea

Secretary's Message

This pupil's book is part of a new Mathematics Program designed and written for use during the early years of schooling in Papua New Guinea. The core materials consist of two pupil books called **Community School Mathematics 2A** and **Community School Mathematics 2B**. They are accompanied by the **Teacher's Resource Book 2**. They will replace the MACS series now in use.

In the Community School Mathematics Program children are taught mathematics by first using real objects. Later, the children will use pictures of objects. Finally, the children will use number symbols to represent these objects. Each school will be supplied with a set of **linking cubes** and a set of **pattern blocks** to use during mathematics lessons. You and your pupils must also collect other real objects such as seeds, nuts, shells etc. **Always allow your pupils to use real objects during their mathematics lessons if they want to. The children will decide when they do not need the help of real objects any longer**. Remember, when children use real objects, they will understand mathematics better.

This program also encourages the children to solve their own problems and make their own decisions with confidence. In order to learn these skills, the children should talk about what they are doing in every lesson. **Since learning is most effective when it has meaning and is enjoyable, allow the children to use the language that they are most comfortable with**. It is the policy of the Department of Education to encourage the use of the vernacular during the early years of schooling, where appropriate. Encourage your pupils and pay attention to what they are saying. If the children practise these skills now, they will be better able to solve problems and make decisions when they become adults.

Finally, the National Department of Education wants teachers to be **flexible** in programming and timetabling. This book will show you some ways to do this.

J. E. Tetaga OBE
Secretary for Education

Contents

Unit 11 *Some Mathematical Shortcuts, Capacity and Angles*

Unit 12 *More on Fractions and Time*

Unit 13 *Revision Activities*

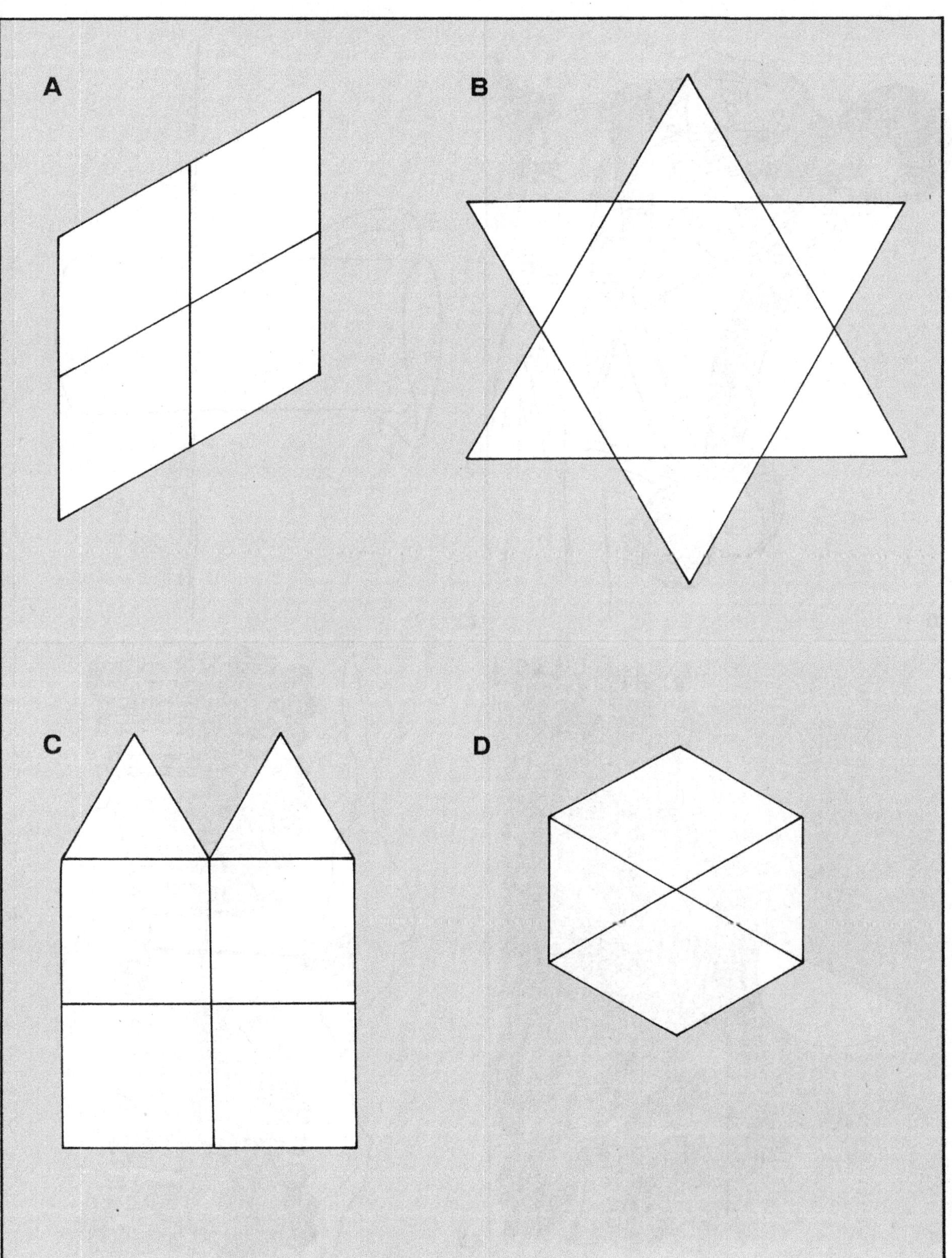
A
B
C
D

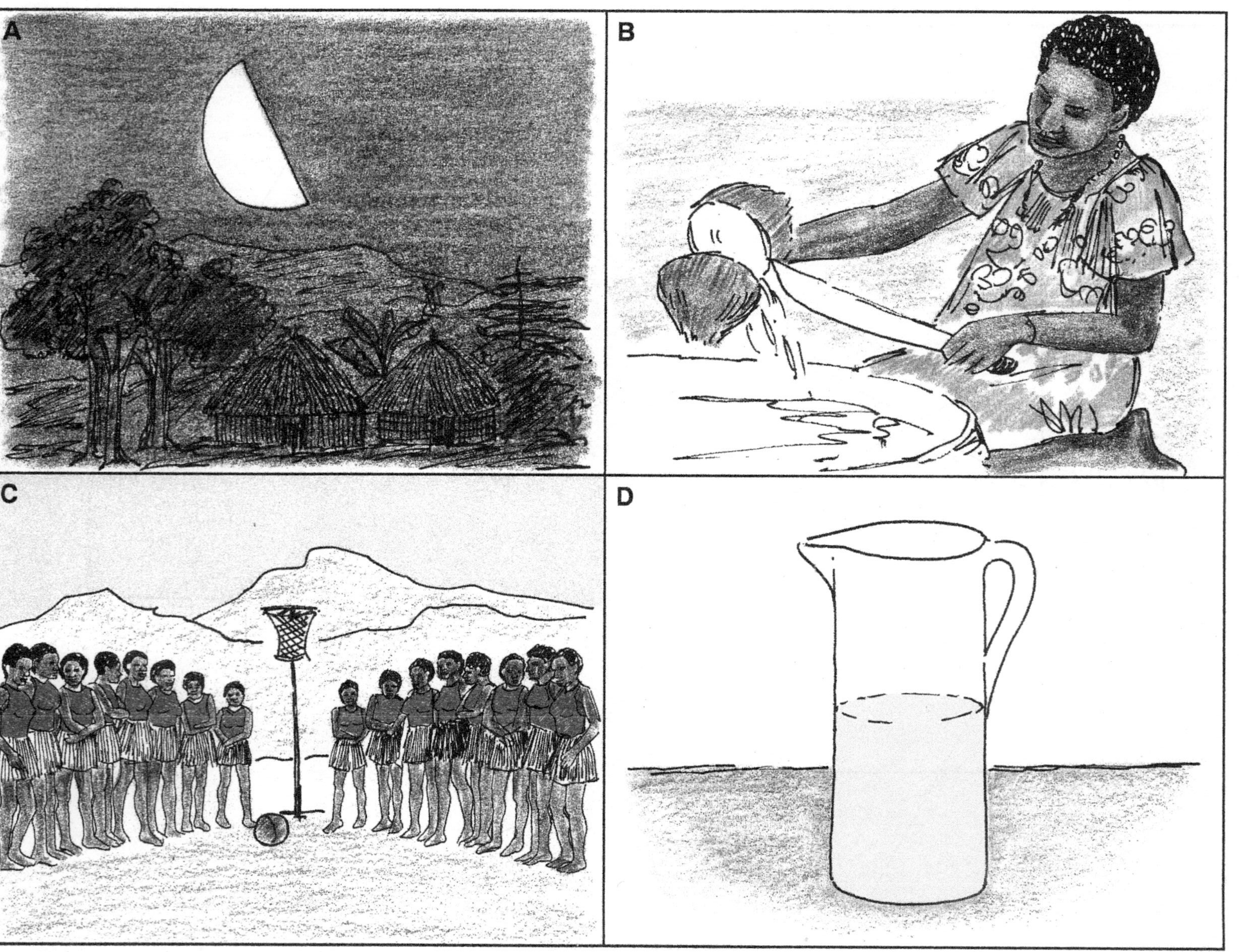
A
B
C
D

A
B

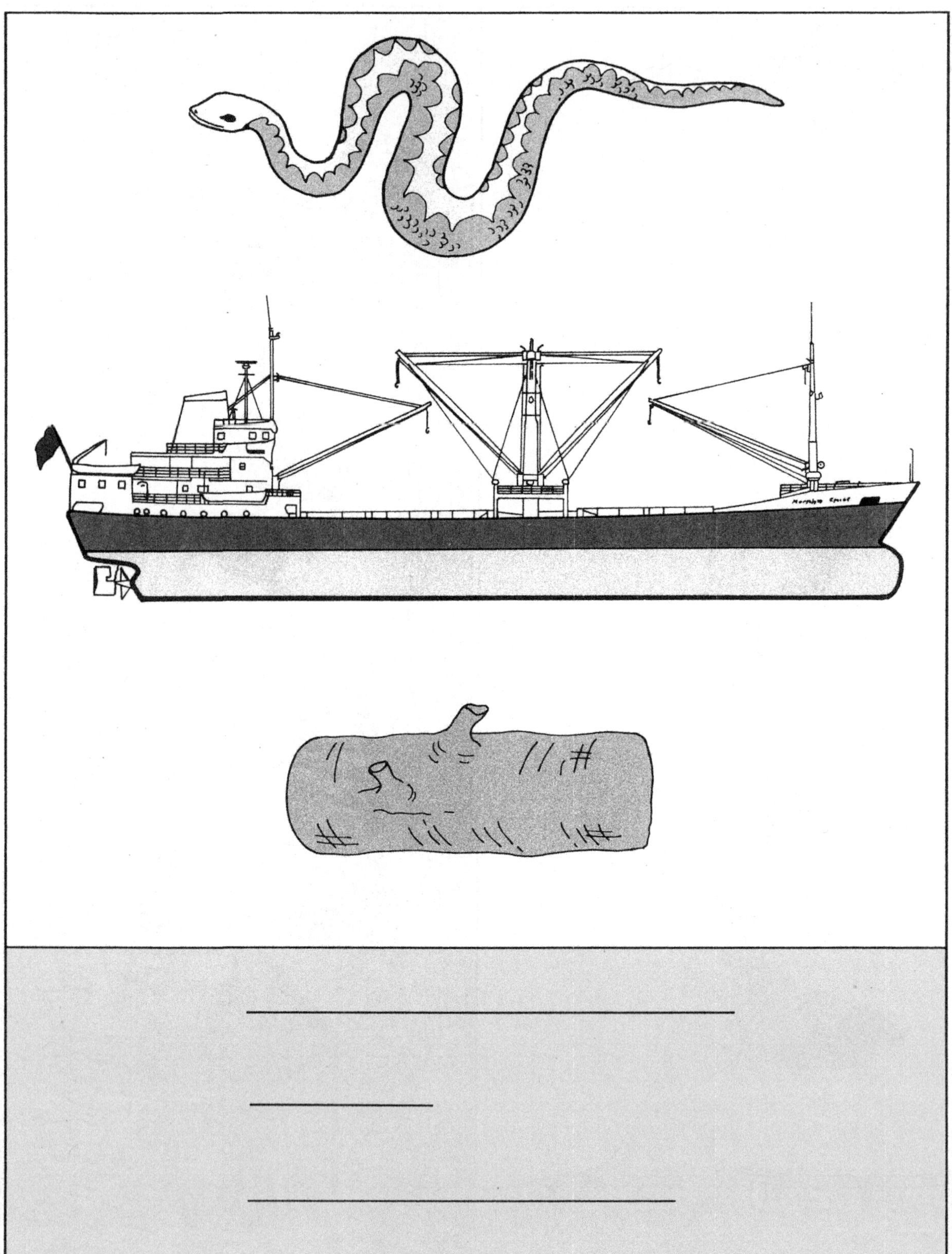

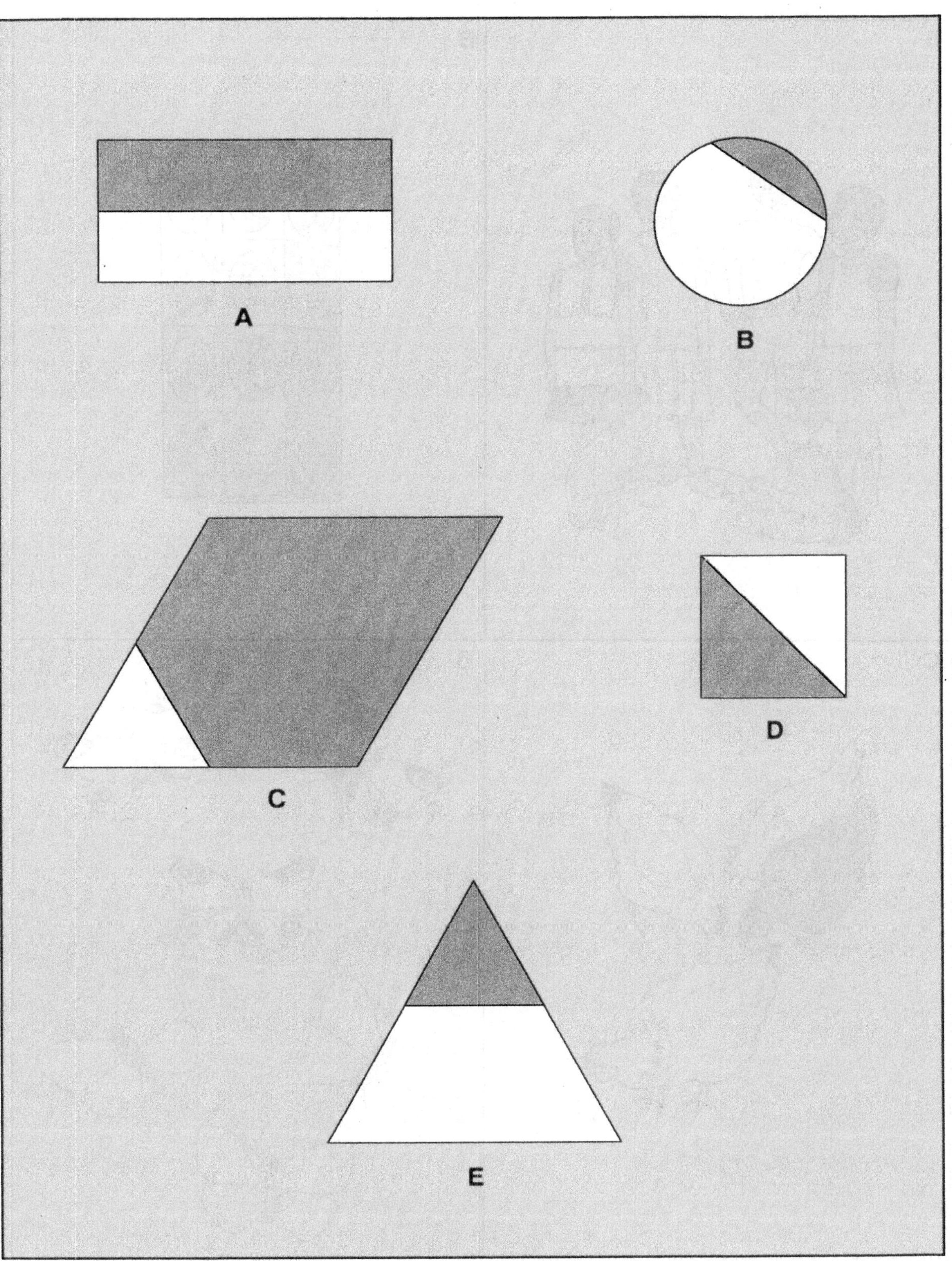
A
B
C
D
E

A

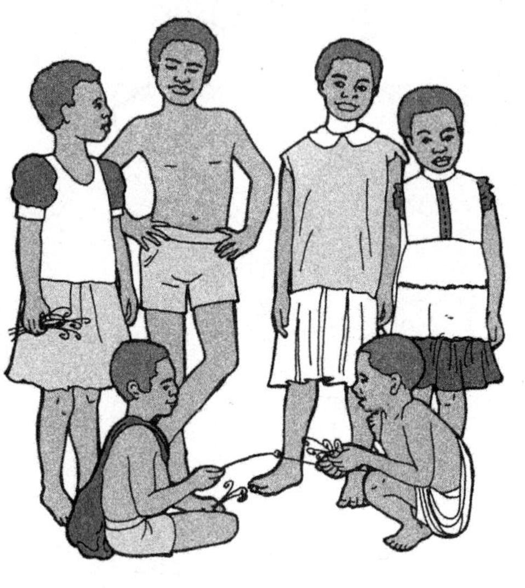

B

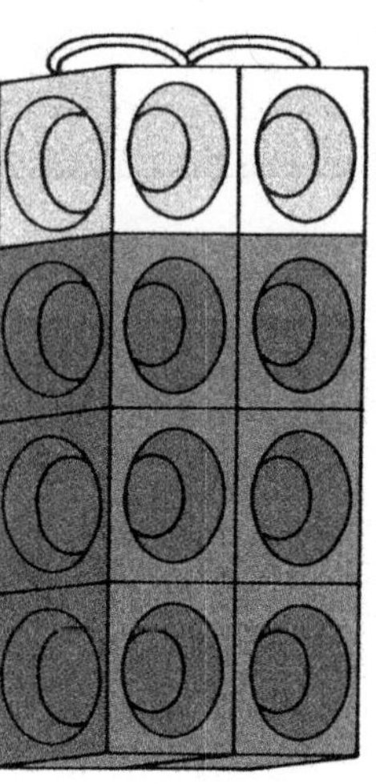

C

D

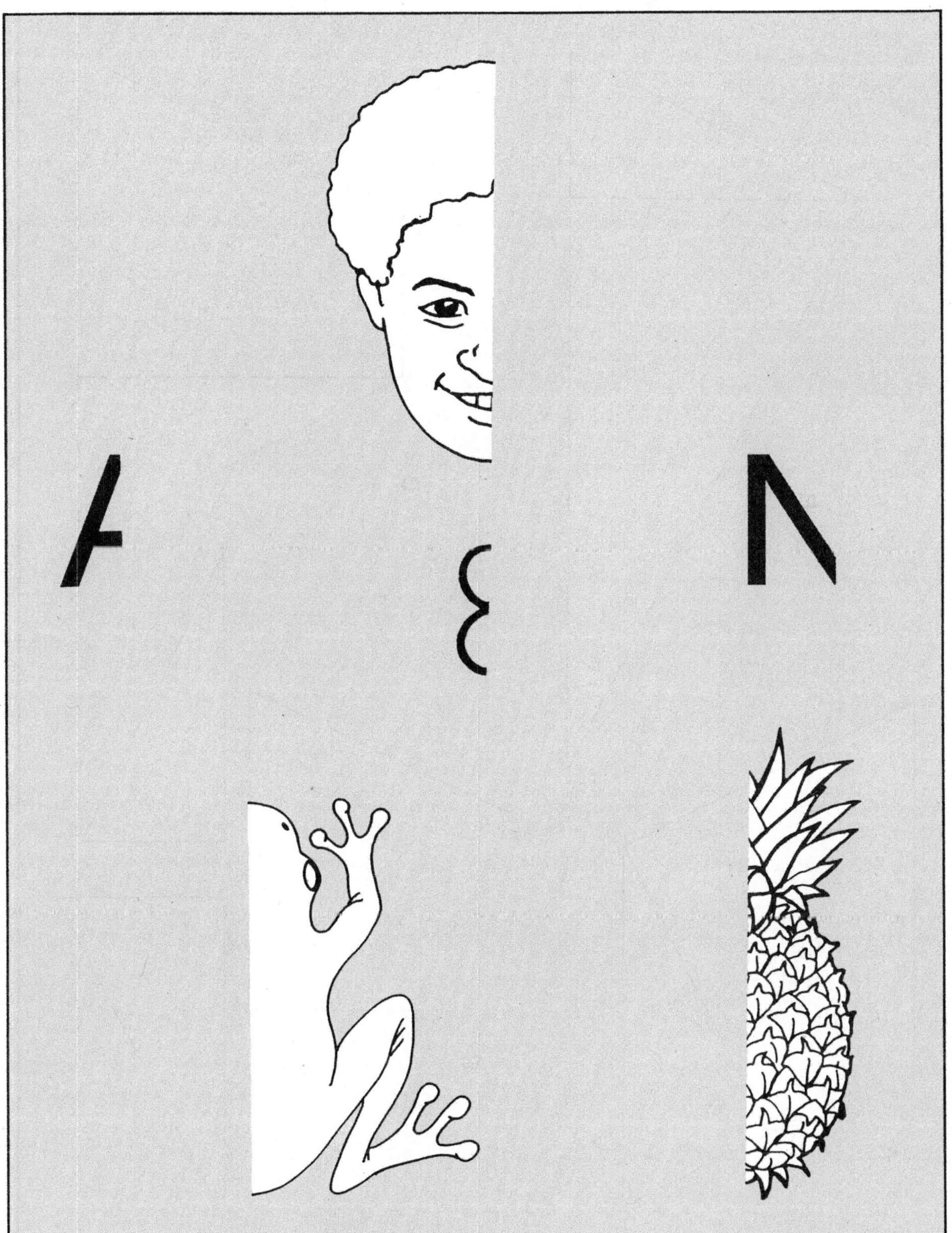

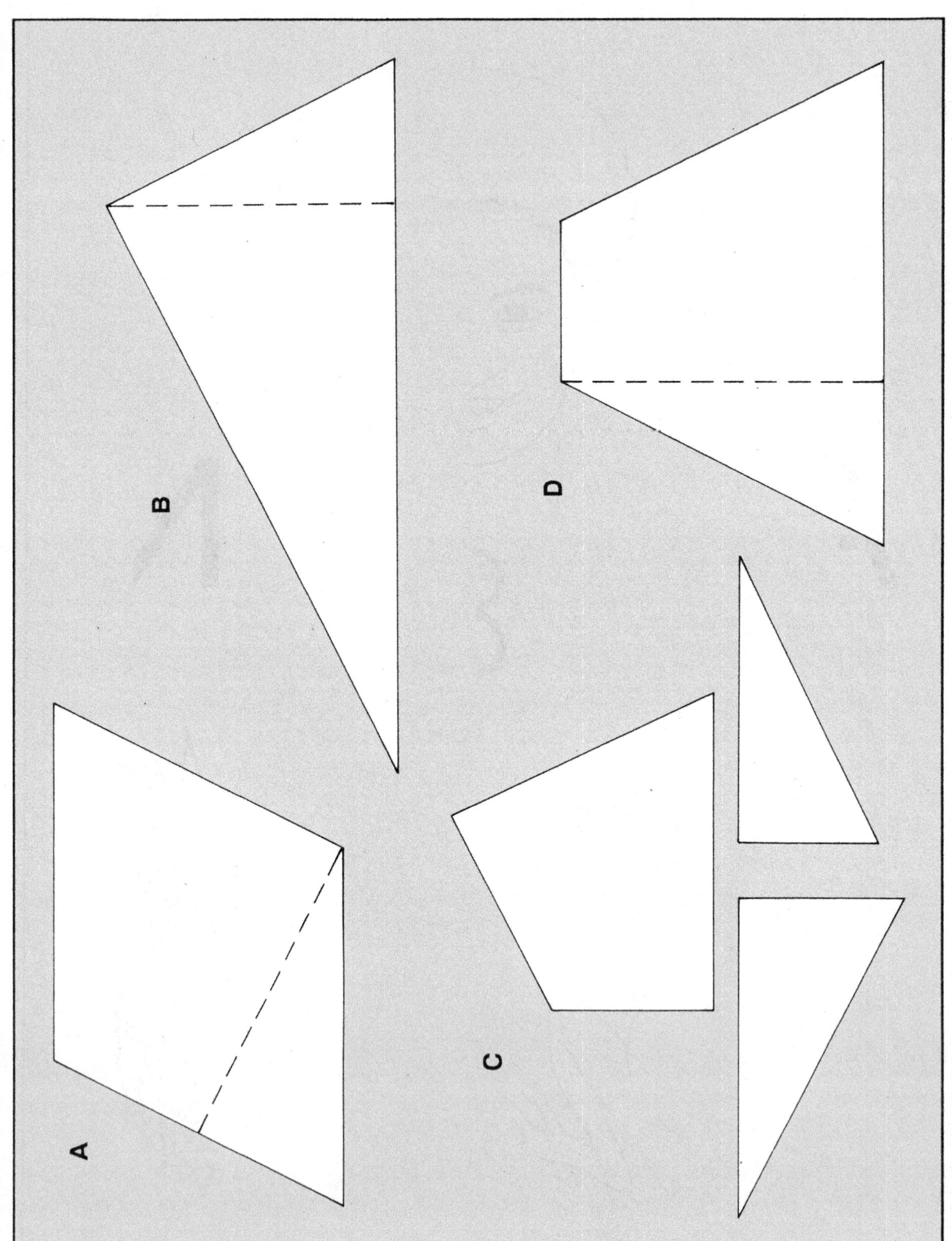
A
B
C
D

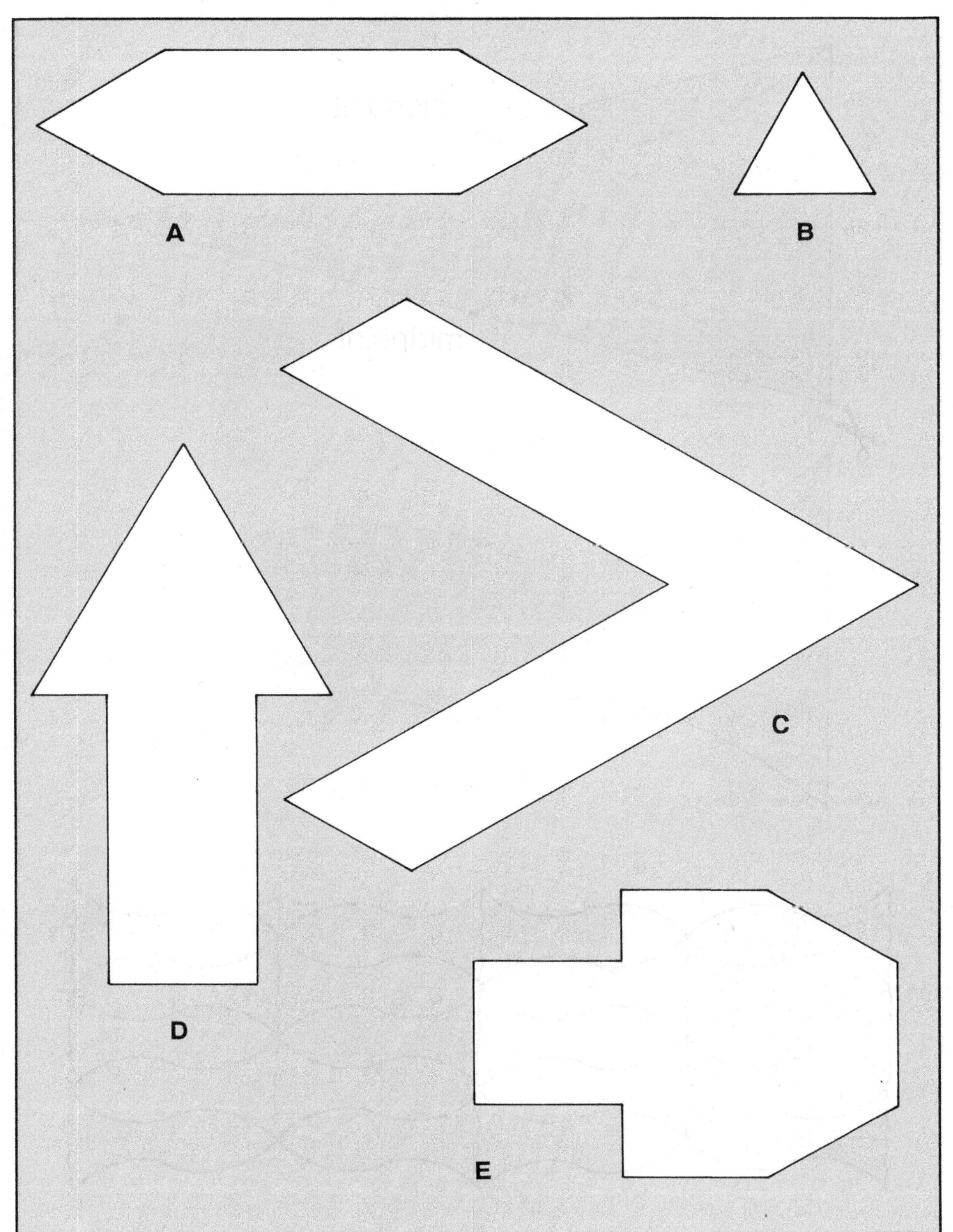
A
B
C
D
E

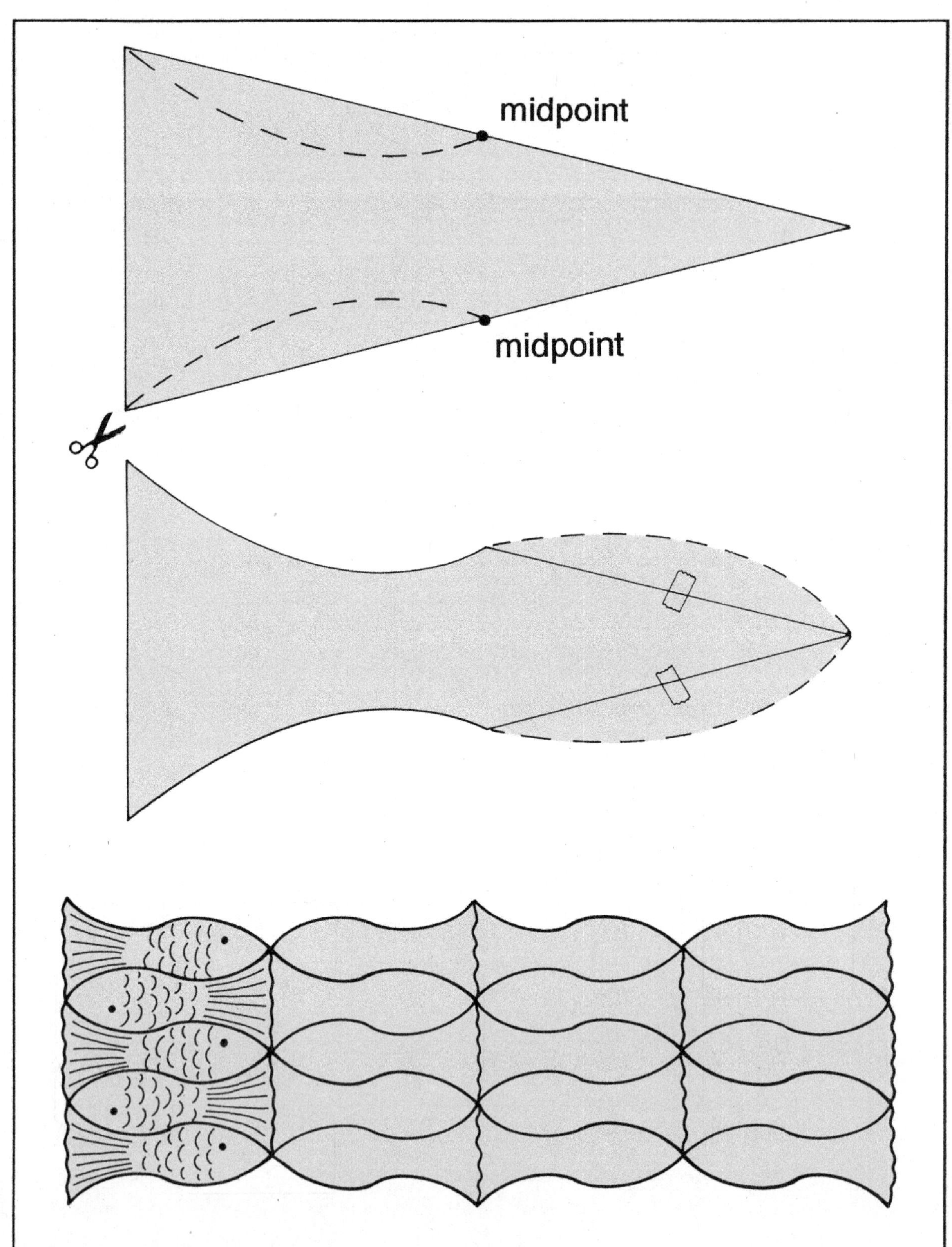
midpoint
midpoint

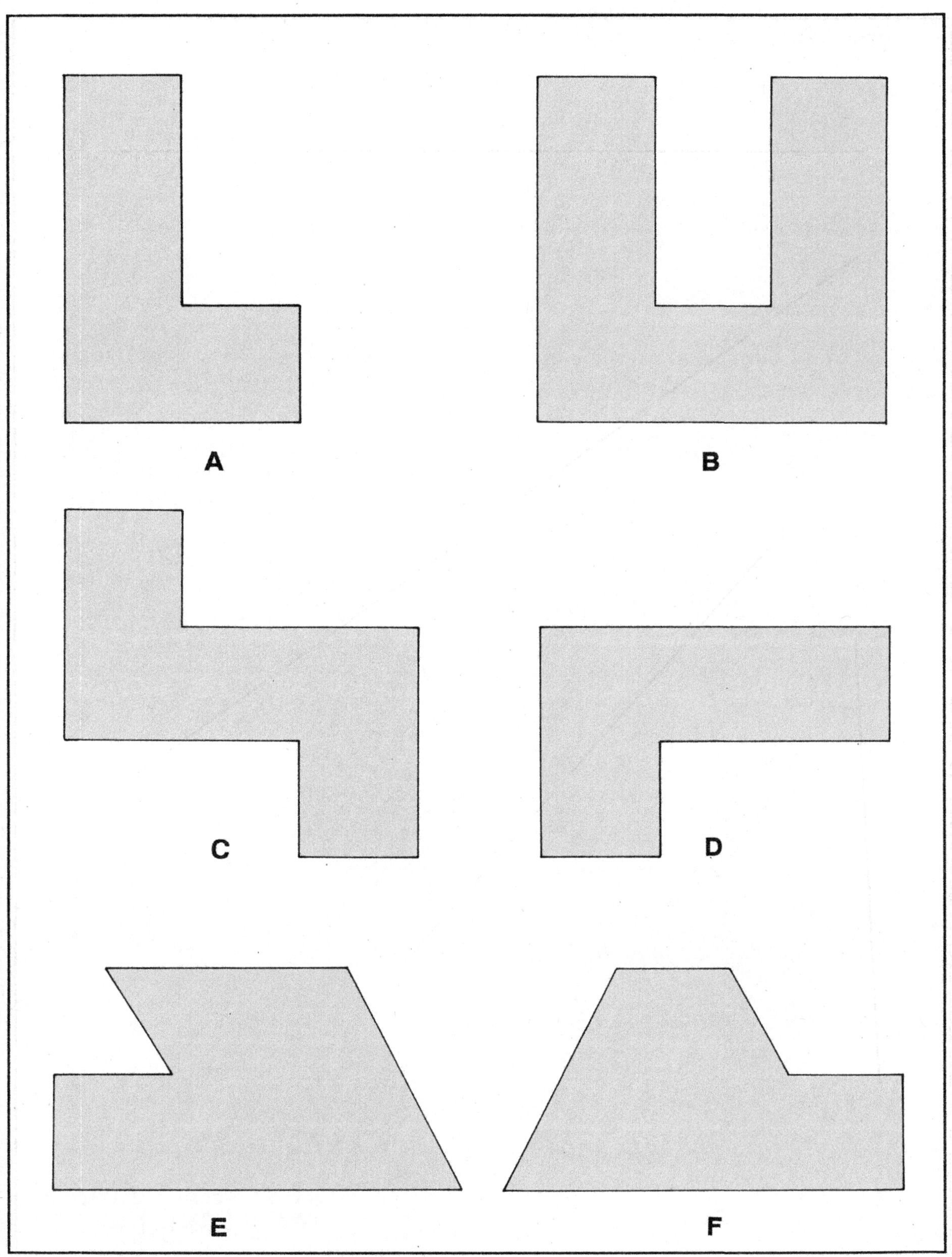
A
B
C
D
E
F

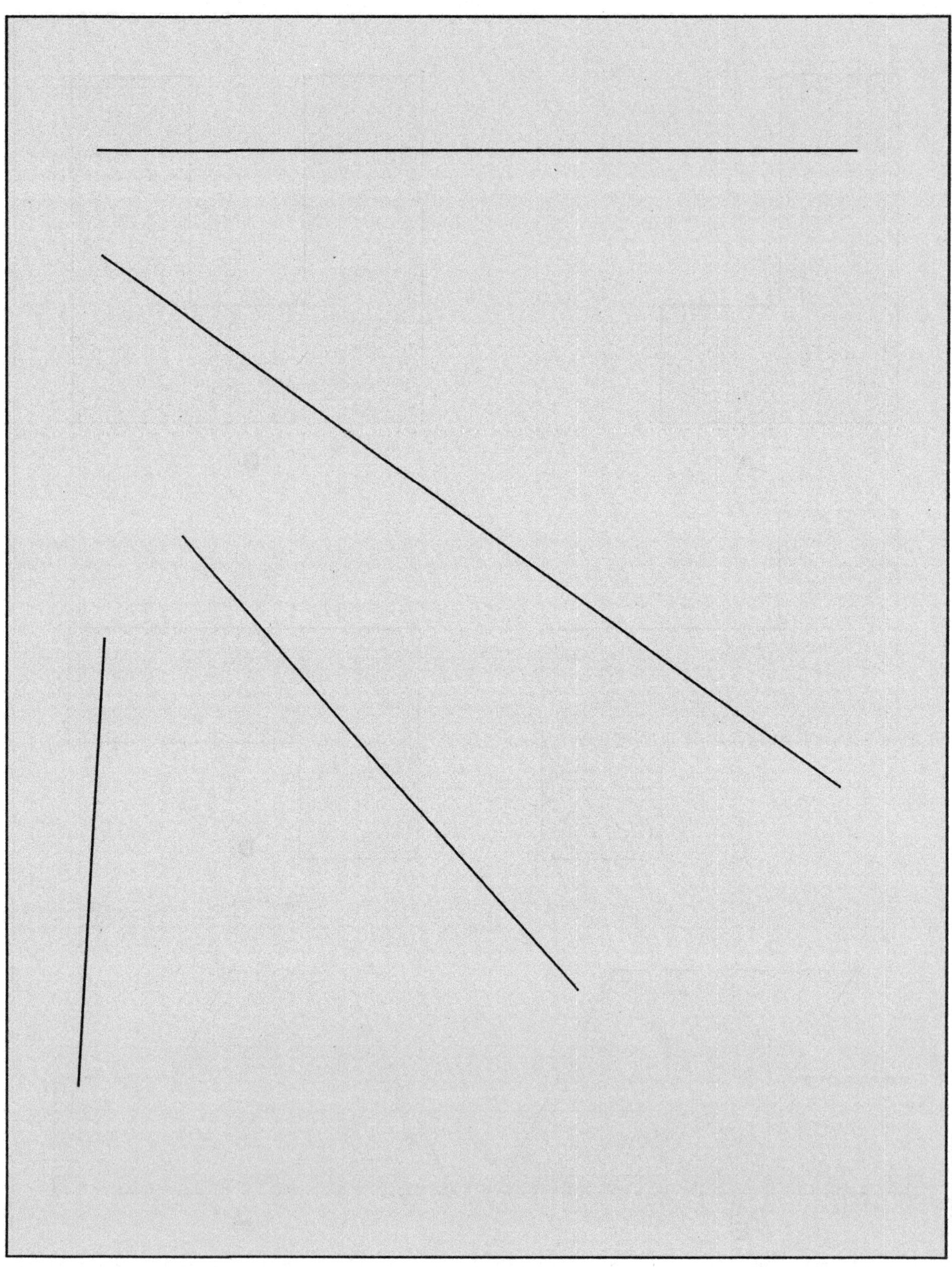

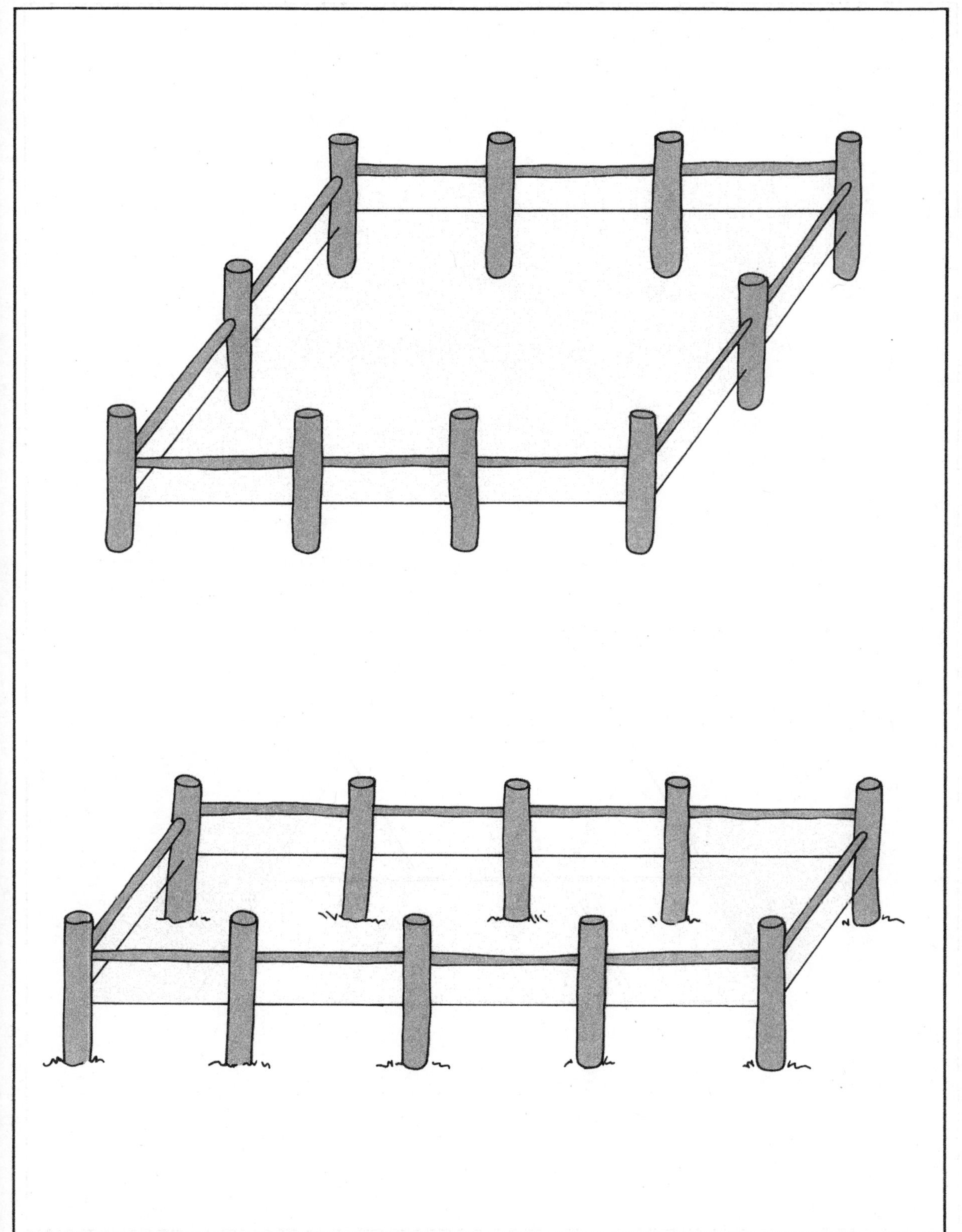

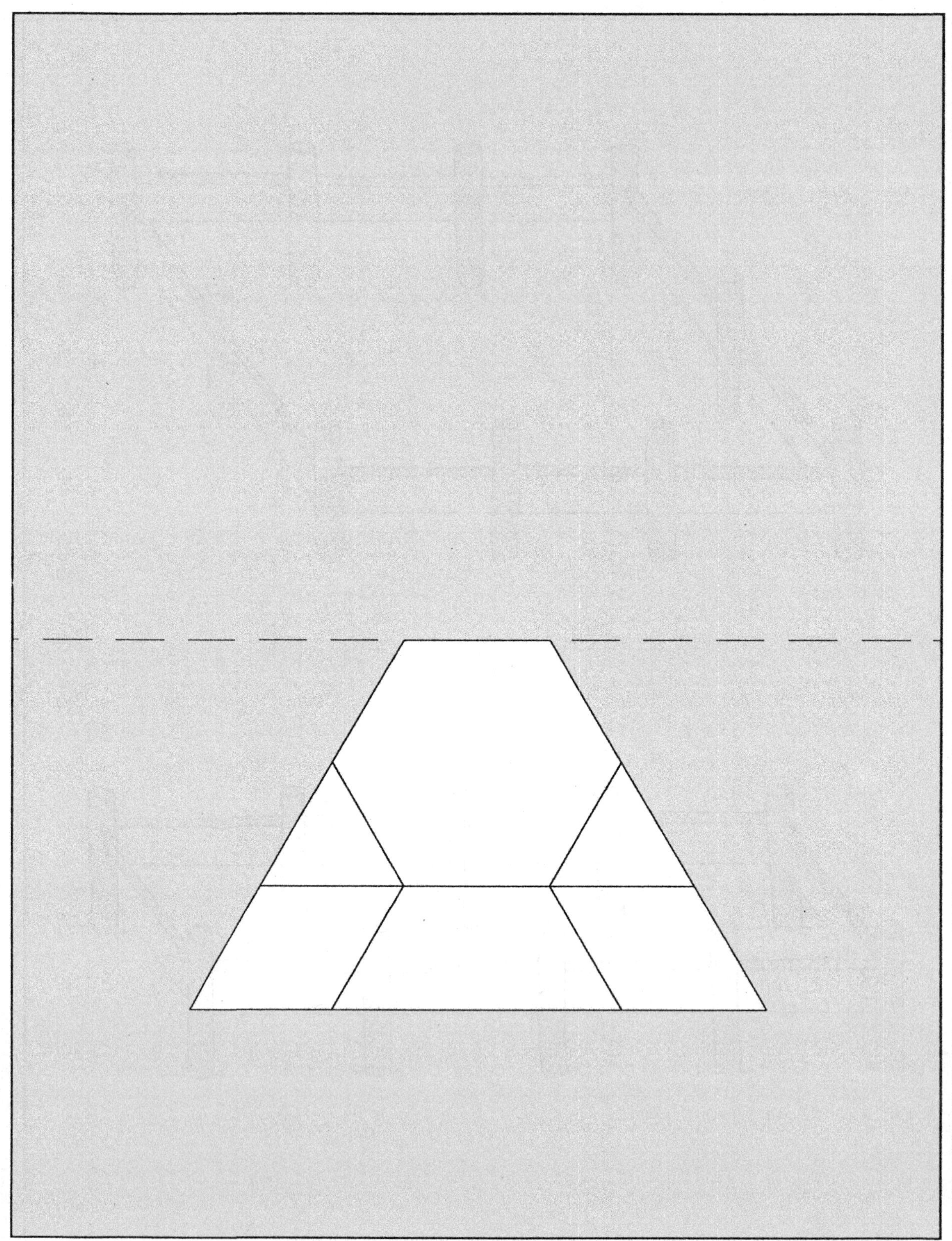

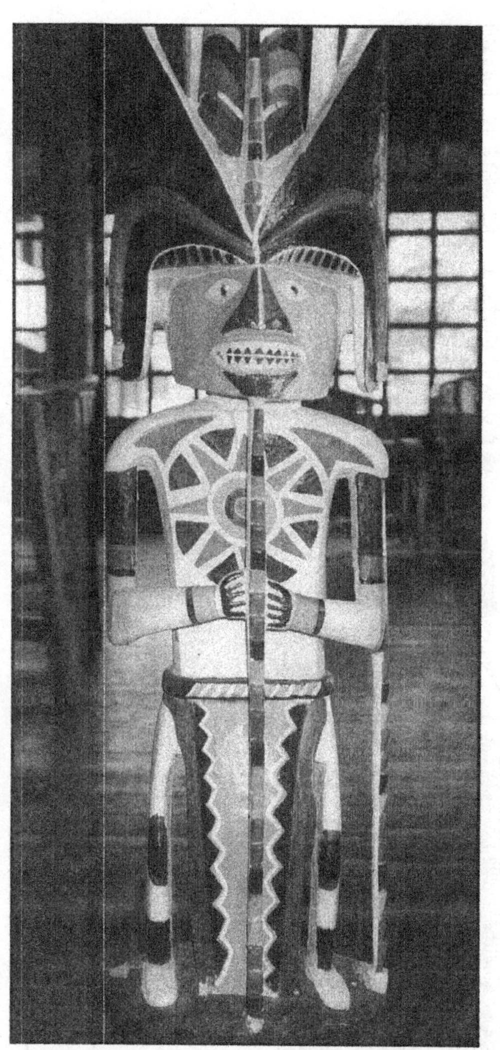

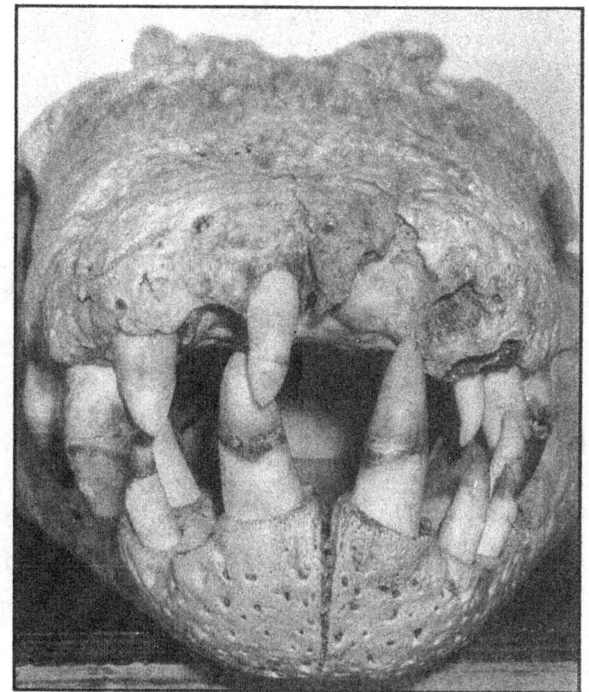

AID POST
STORE

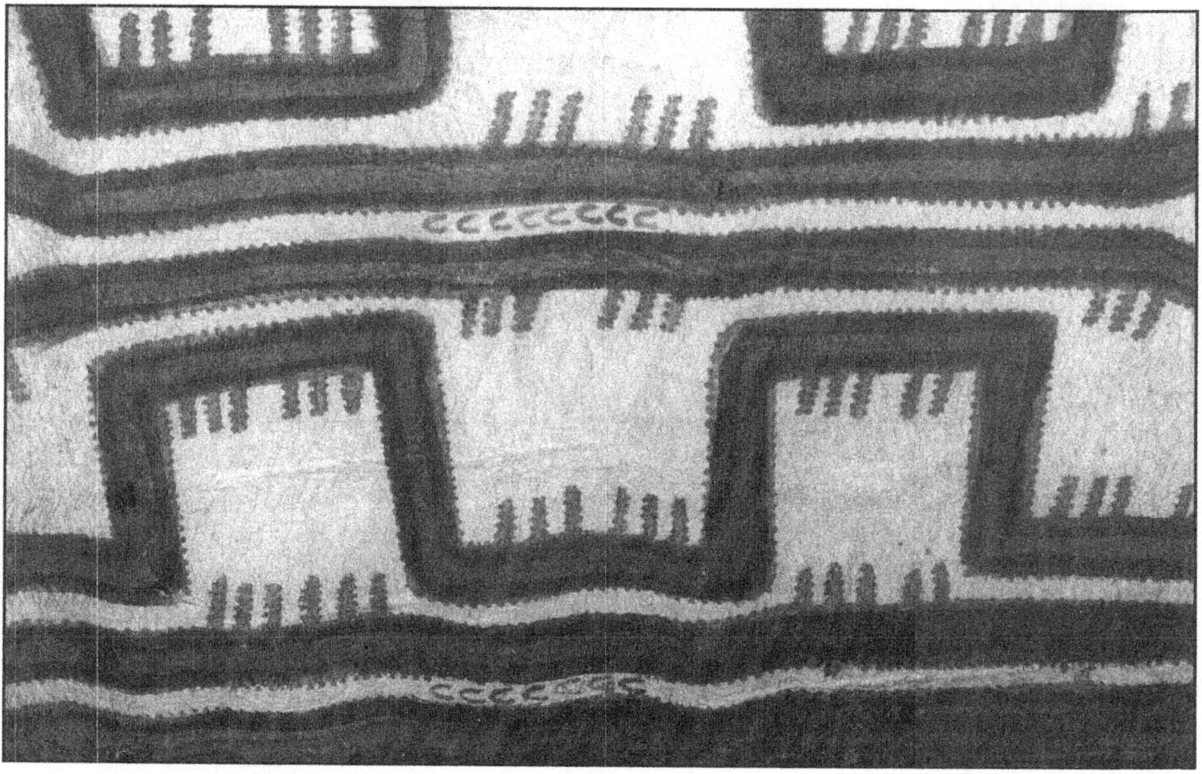

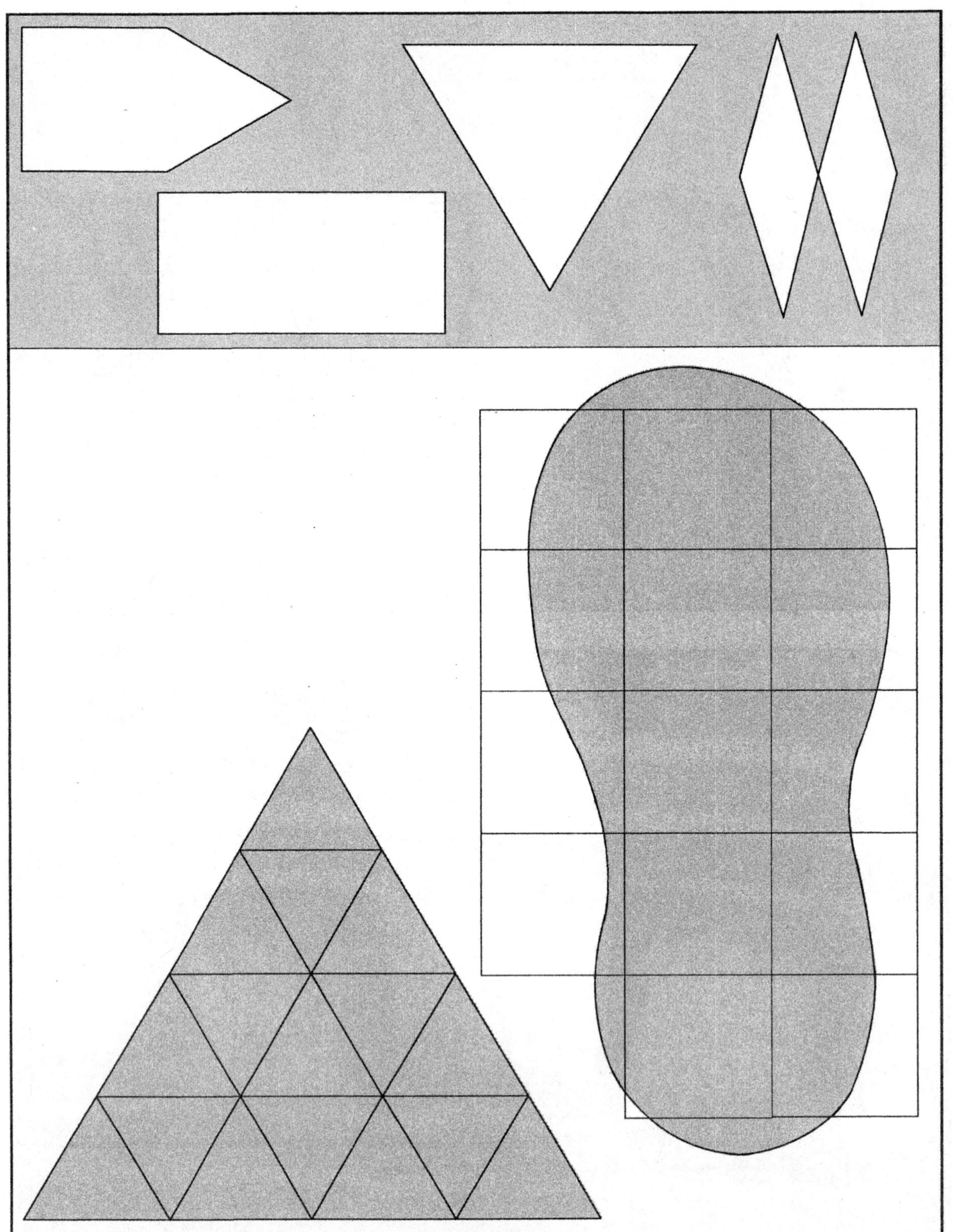

0	10	20	30	40	50	60	70	80	90
100	110	120	130	140	150	160	170	180	190
200	210	220	230	240	250	260	270	280	290
300	310	320	330	340	350	360	370	380	390
400	410	420	430	440	450	460	470	480	490
500	510	520	530	540	550	560	570	580	590
600	610	620	630	640	650	660	670	680	690
700	710	720	730	740	750	760	770	780	790
800	810	820	830	840	850	860	870	880	890
900	910	920	930	940	950	960	970	980	990

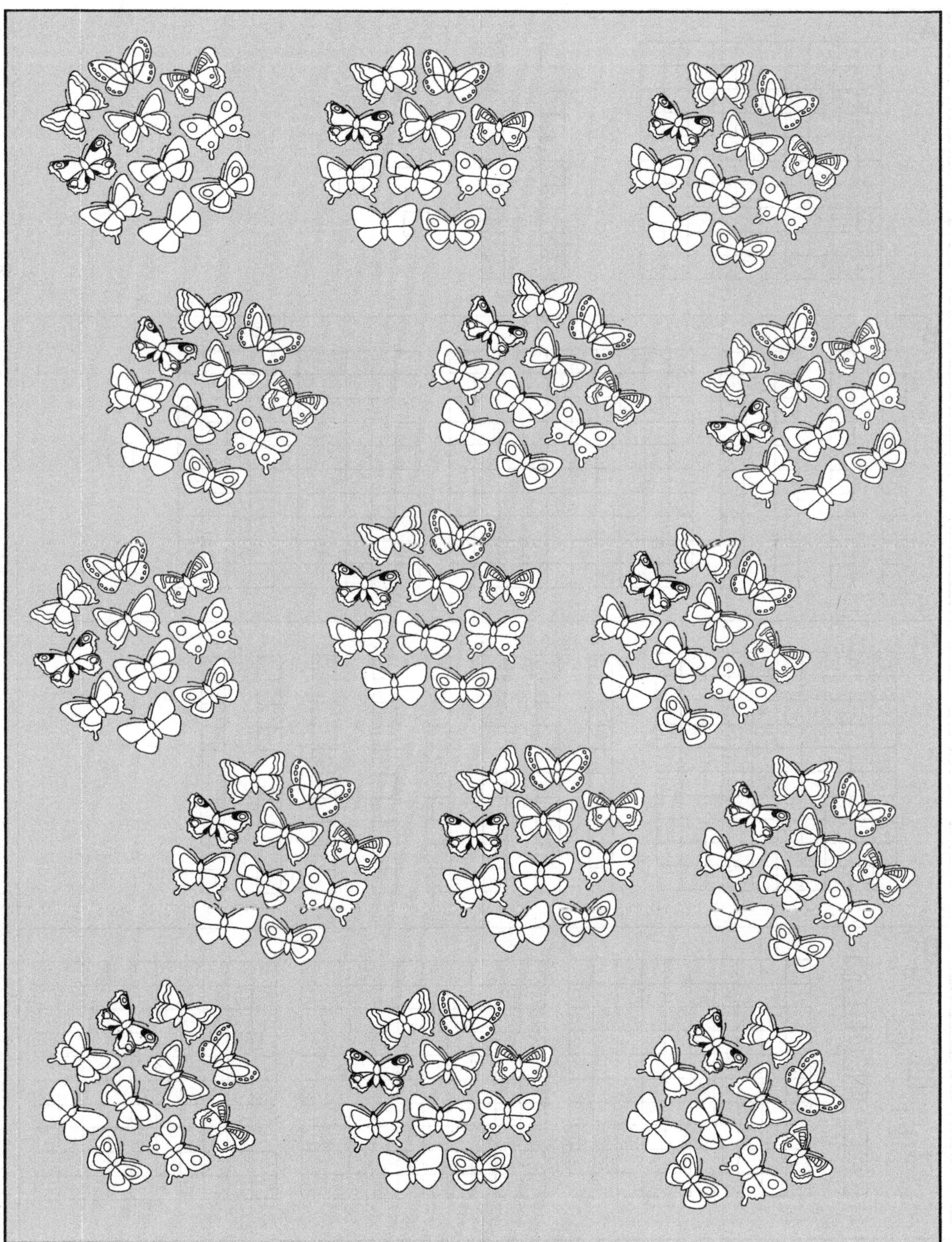

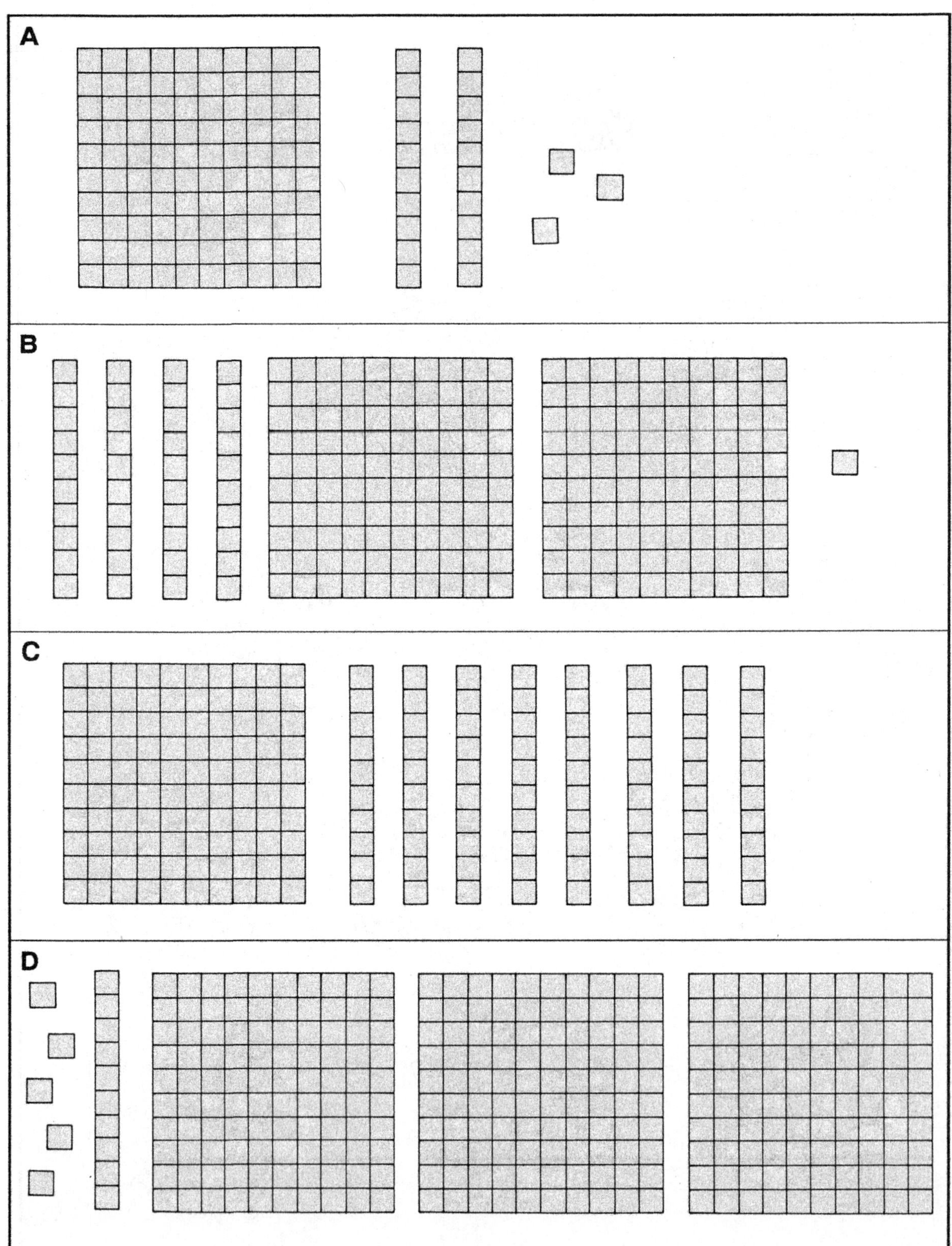
A
B
C
D

A
100
100
10
10
10
1
B
100
100
100
100
10
10
10
10
10
1
1
1
1
1
C
100
100
100
10
10
10
10
10
1
1
1
1
1
1
1
1
1
D
100
10
10
10
10
10
10
10
1
1
1
1

146

538

477

926

231

816

301

222

692

737

502

1	2	3	4	5	6	7	8	9	10
11	12	13	14	15	16	17	18	19	20
21	22	23	24	25	26	27	28	29	30
31	32	33	34	35	36	37	38	39	40
41	42	43	44	45	46	47	48	49	50
51	52	53	54	55	56	57	58	59	60
61	62	63	64	65	66	67	68	69	70
71	72	73	74	75	76	77	78	79	80
81	82	83	84	85	86	87	88	89	90
91	92	93	94	95	96	97	98	99	100
101	102	103	104	105	106	107	108	109	110
111	112	113	114	115	116	117	118	119	120
121	122	123	124	125	126	127	128	129	130
131	132	133	134	135	136	137	138	139	140
141	142	143	144	145	146	147	148	149	150
151	152	153	154	155	156	157	158	159	160
161	162	163	164	165	166	167	168	169	170
171	172	173	174	175	176	177	178	179	180
181	182	183	184	185	186	187	188	189	190
191	192	193	194	195	196	197	198	199	200

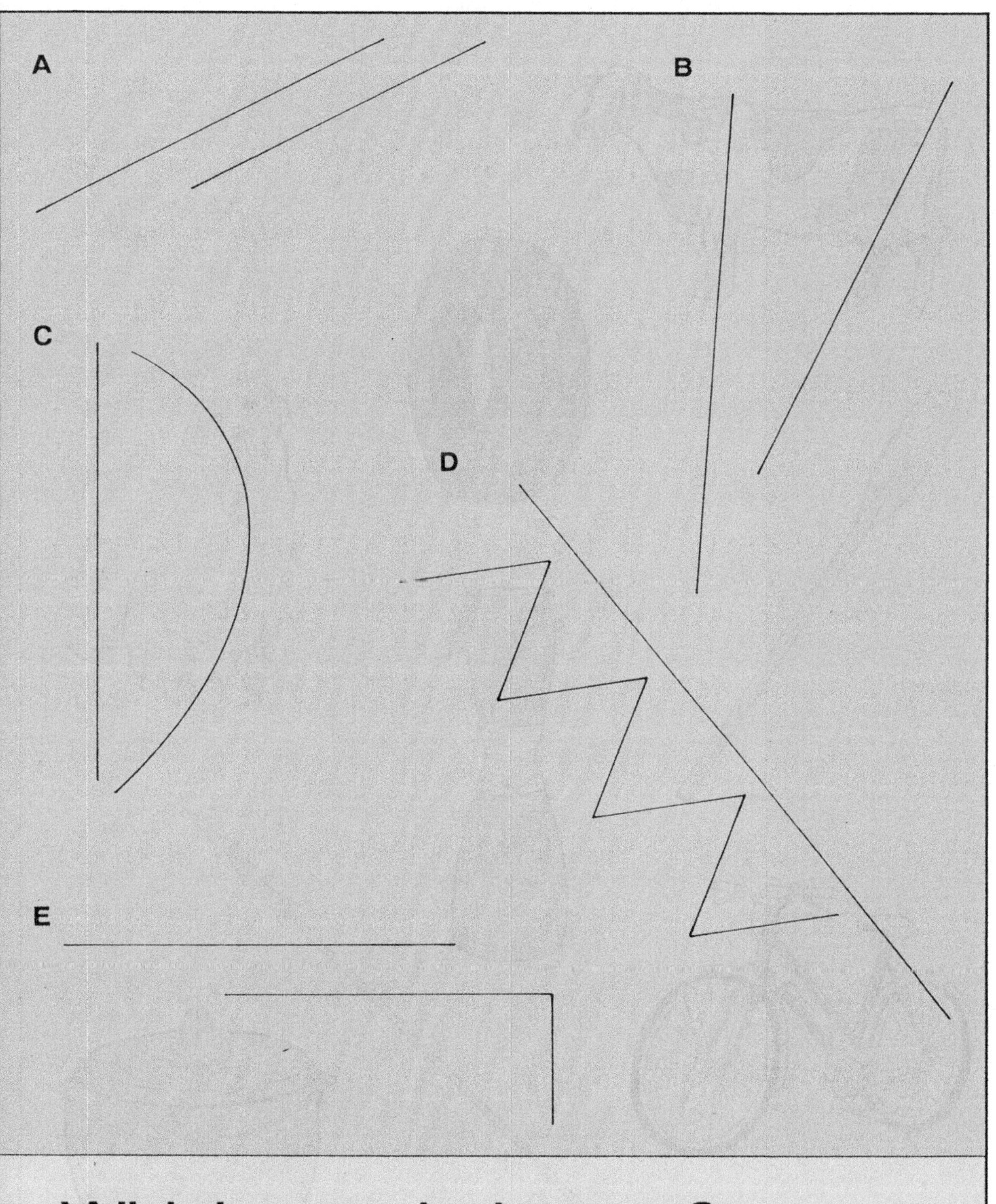

Which one is longer?

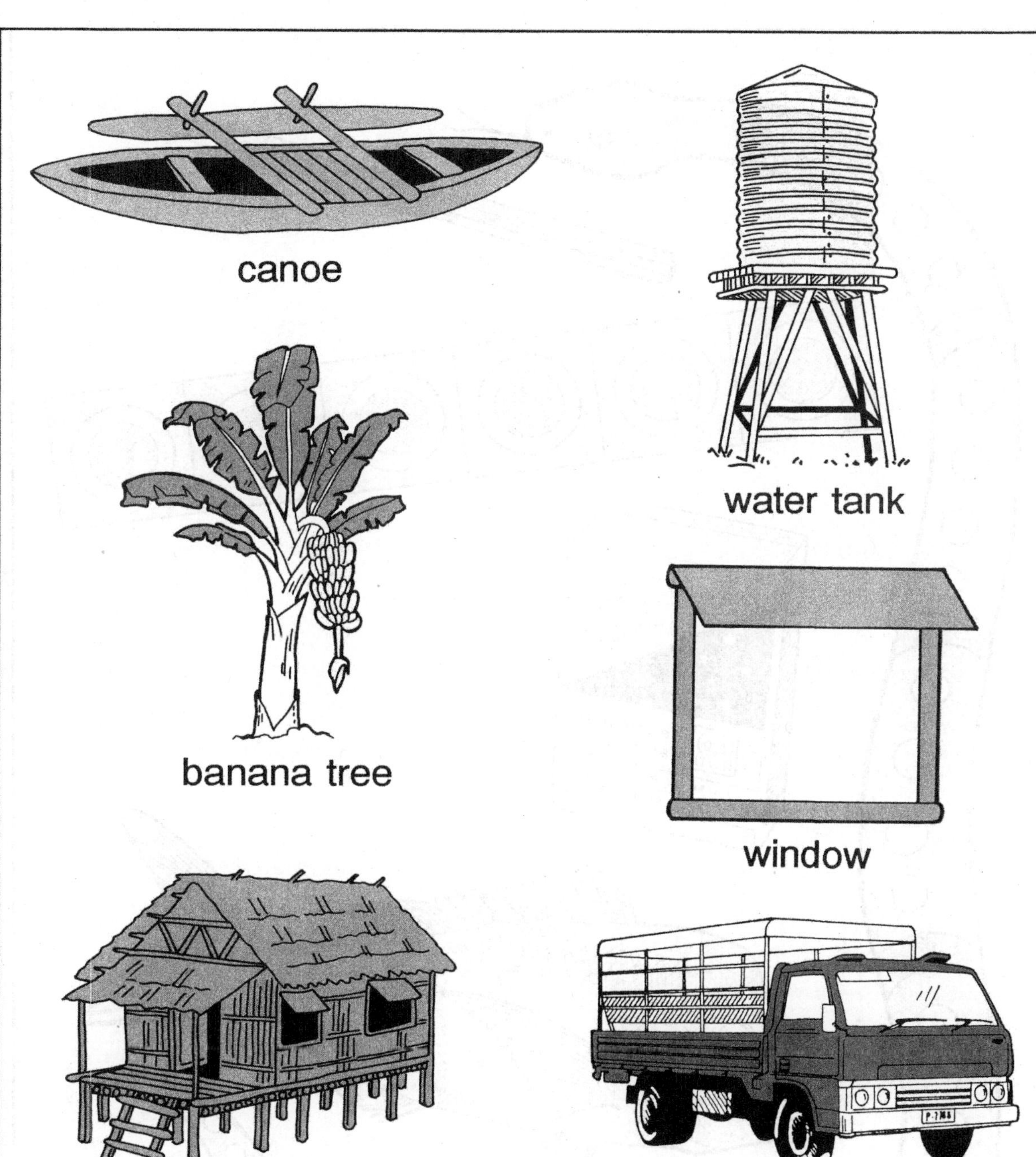

How long are these?

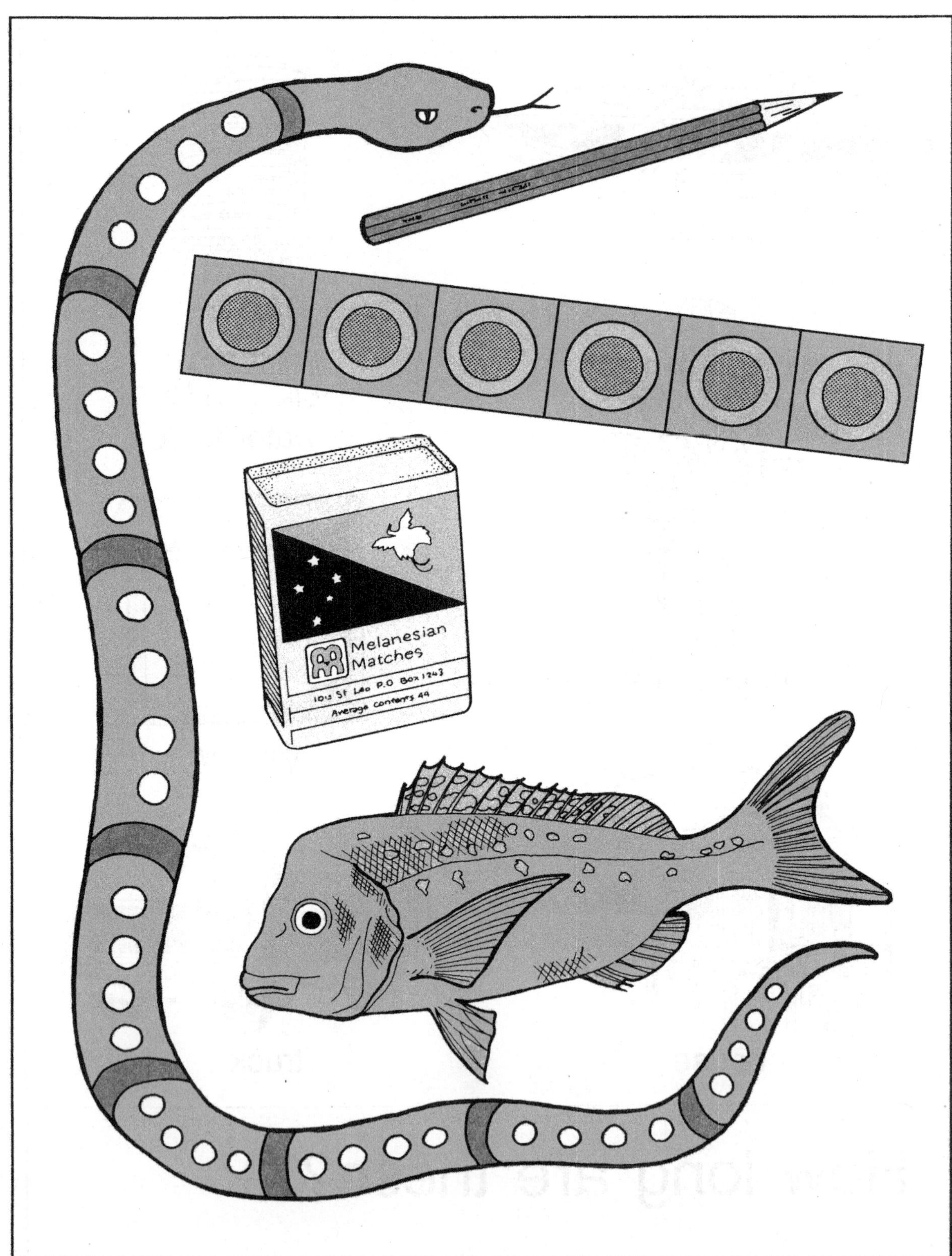
Melanesian
Matches
1013 St Leo P.O Box 1243
Average contents 44

$$\begin{array}{r} 34 \\ +\,22 \\ \hline \end{array} \quad \begin{array}{r} 35 \\ +\,27 \\ \hline \end{array} \quad \begin{array}{r} 46 \\ +\,35 \\ \hline \end{array} \quad \begin{array}{r} 33 \\ +\,79 \\ \hline \end{array}$$

$$\begin{array}{r} 41 \\ +\,56 \\ \hline \end{array} \quad \begin{array}{r} 29 \\ +\,74 \\ \hline \end{array} \quad \begin{array}{r} 27 \\ +\,37 \\ \hline \end{array} \quad \begin{array}{r} 32 \\ +\,86 \\ \hline \end{array}$$

Make up a story.

$$\begin{array}{r} 32 \\ +\,41 \\ \hline \end{array}$$

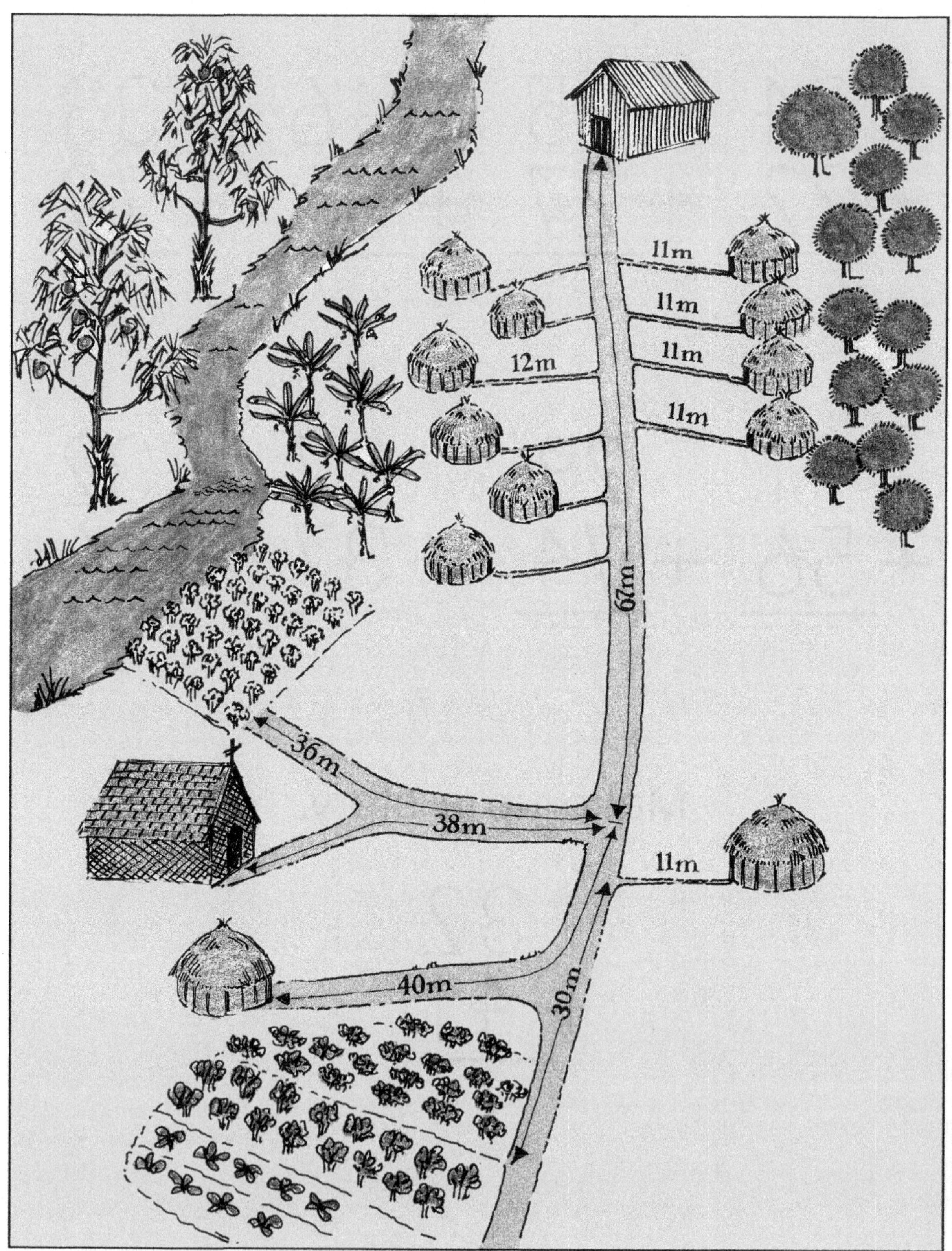
11m
11m
11m
12m
11m
67m
36m
38m
11m
40m
30m

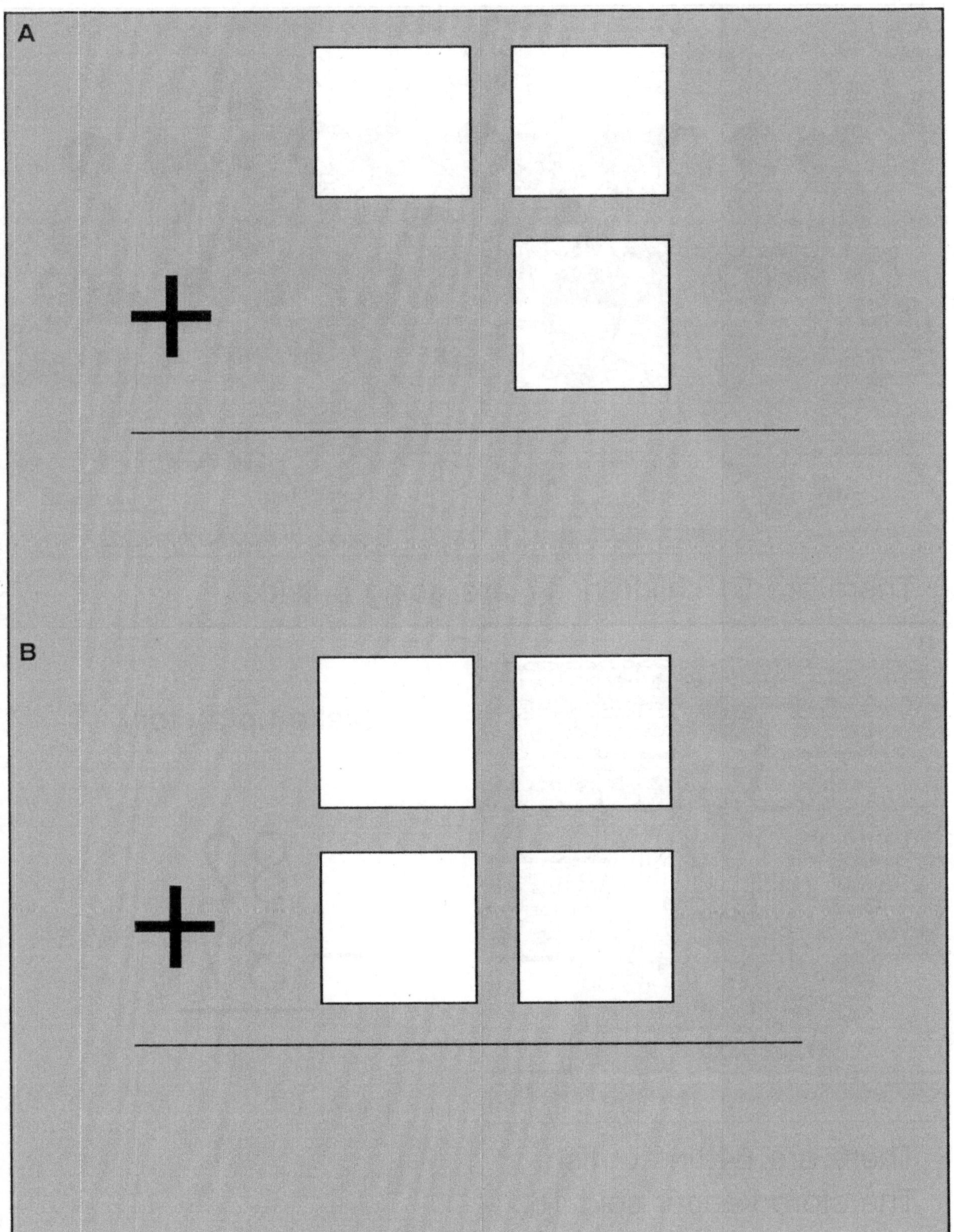
A
B

A

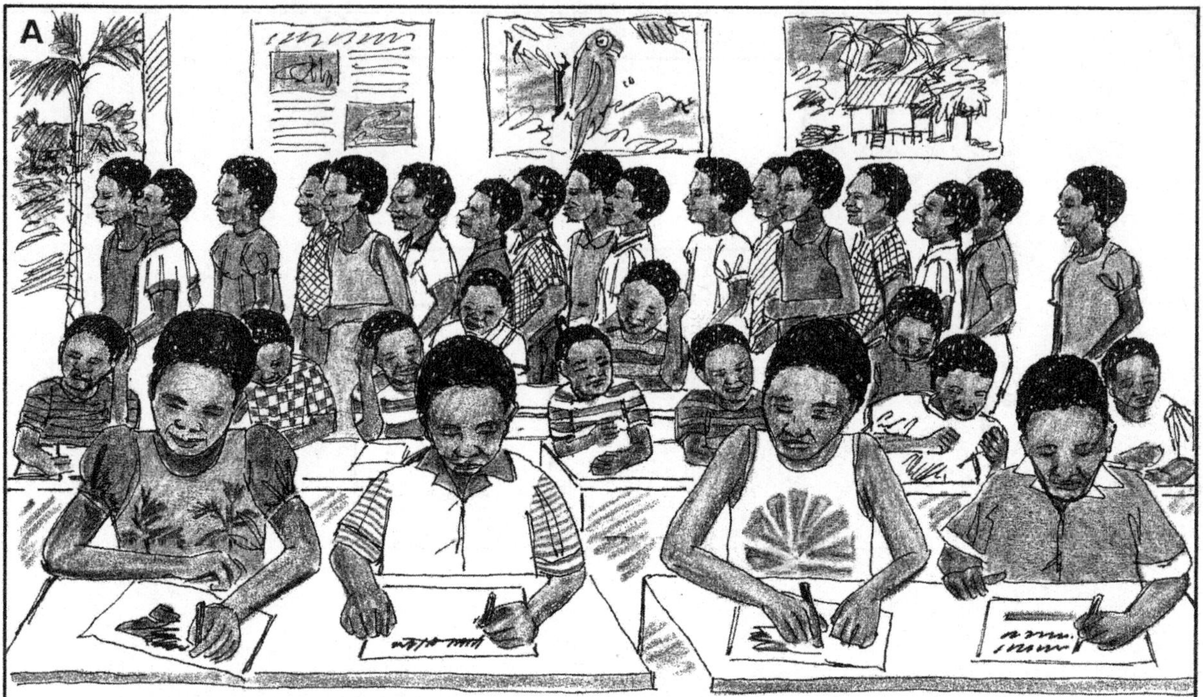

There are 31 children. 17 are going outside.

B

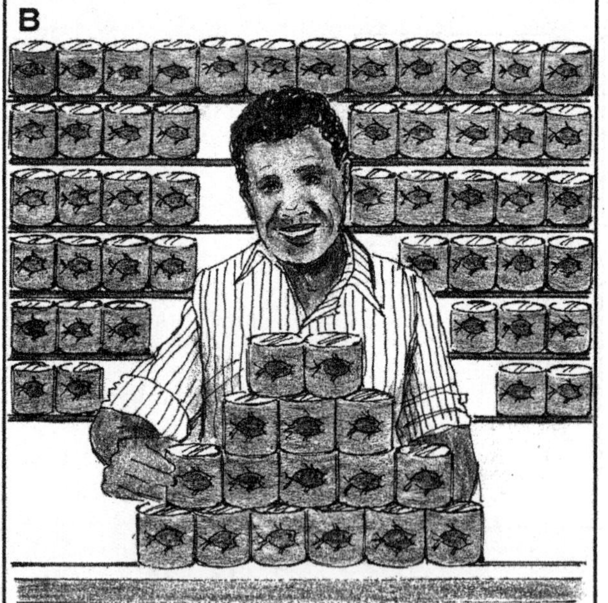

There are 64 tins of fish.
The store keeper sold 16.

C

Make up a story.

$$\begin{array}{r} 82 \\ -\ 37 \\ \hline \end{array}$$

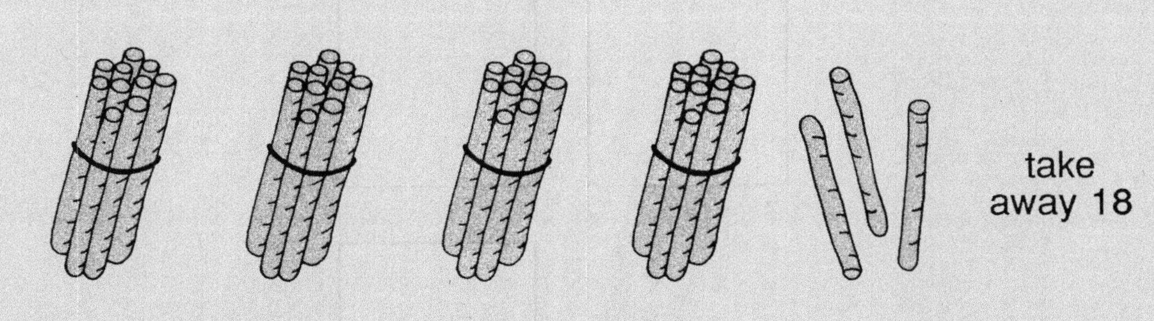

A

$$\begin{array}{r} 52 \\ -\ 35 \\ \hline \end{array}$$

B

$$\begin{array}{r} 38 \\ -\ 12 \\ \hline \end{array}$$

C

$$\begin{array}{r} 45 \\ -\ 21 \\ \hline \end{array}$$

D

$$\begin{array}{r} 48 \\ -\ 29 \\ \hline \end{array}$$

E

$$\begin{array}{r} 82 \\ -\ 56 \\ \hline \end{array}$$

F

$$\begin{array}{r} 44 \\ -\ 22 \\ \hline \end{array}$$

G

$$\begin{array}{r} 34 \\ -\ 16 \\ \hline \end{array}$$

H

$$\begin{array}{r} 81 \\ -\ 17 \\ \hline \end{array}$$

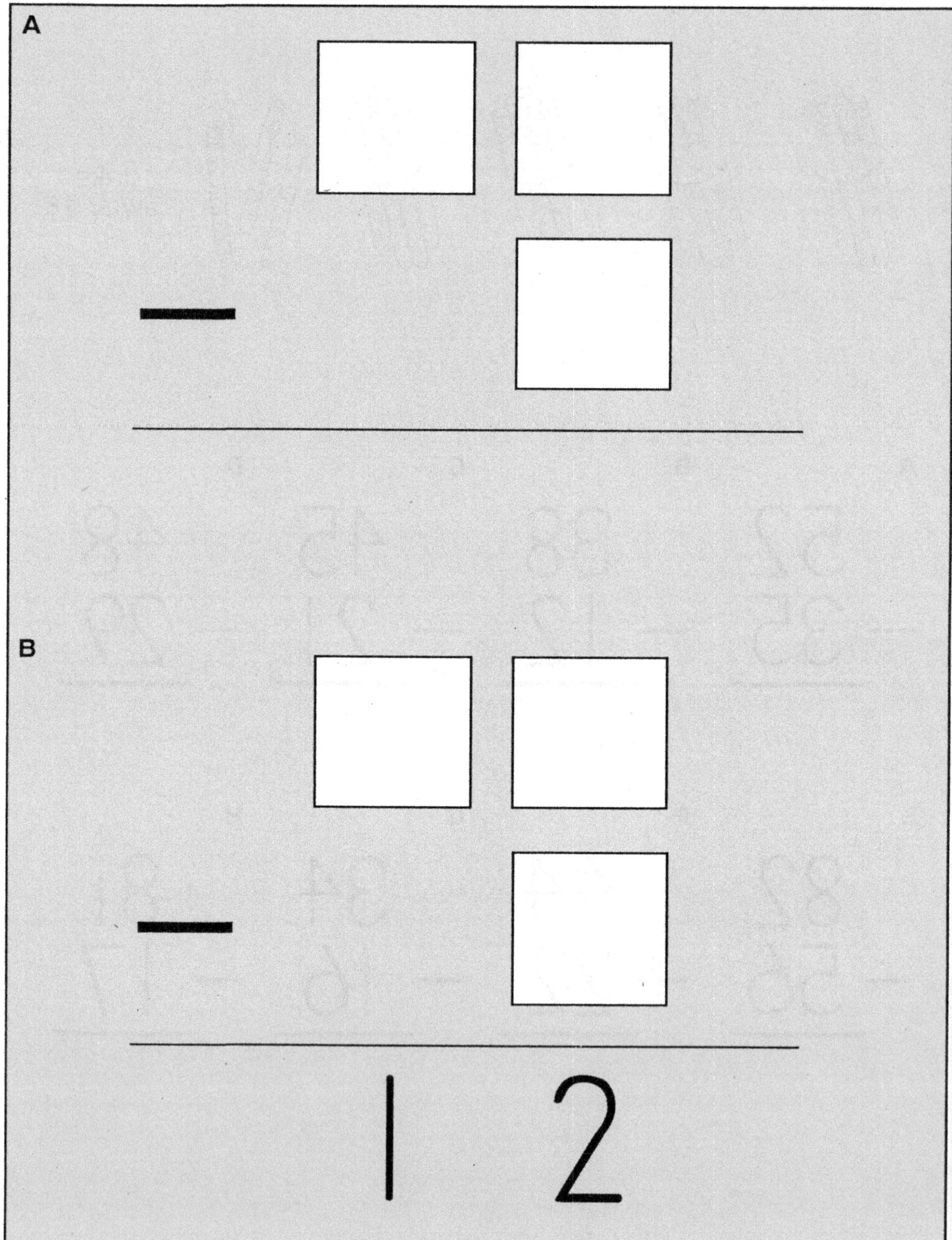
A
B
1 2

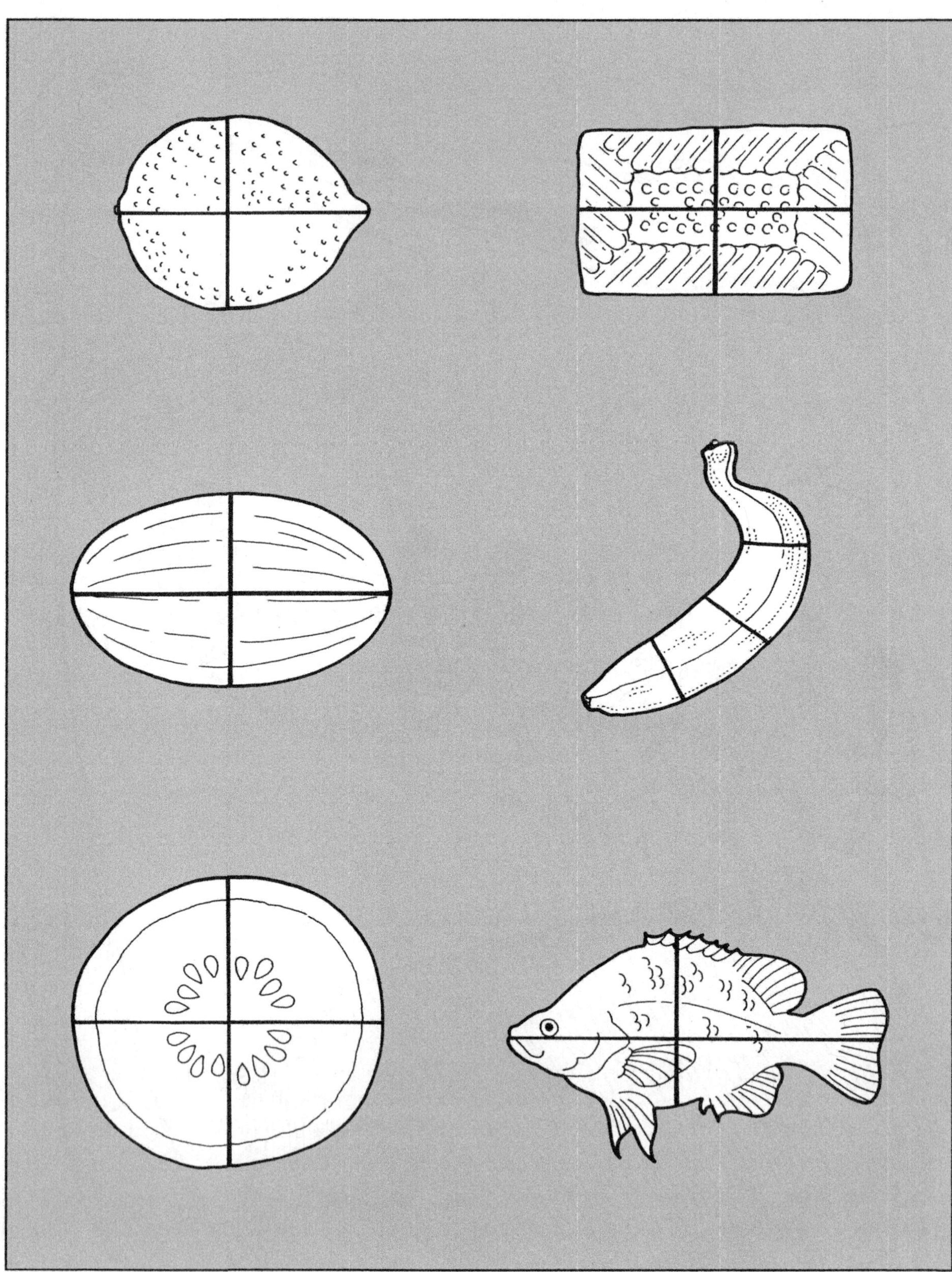

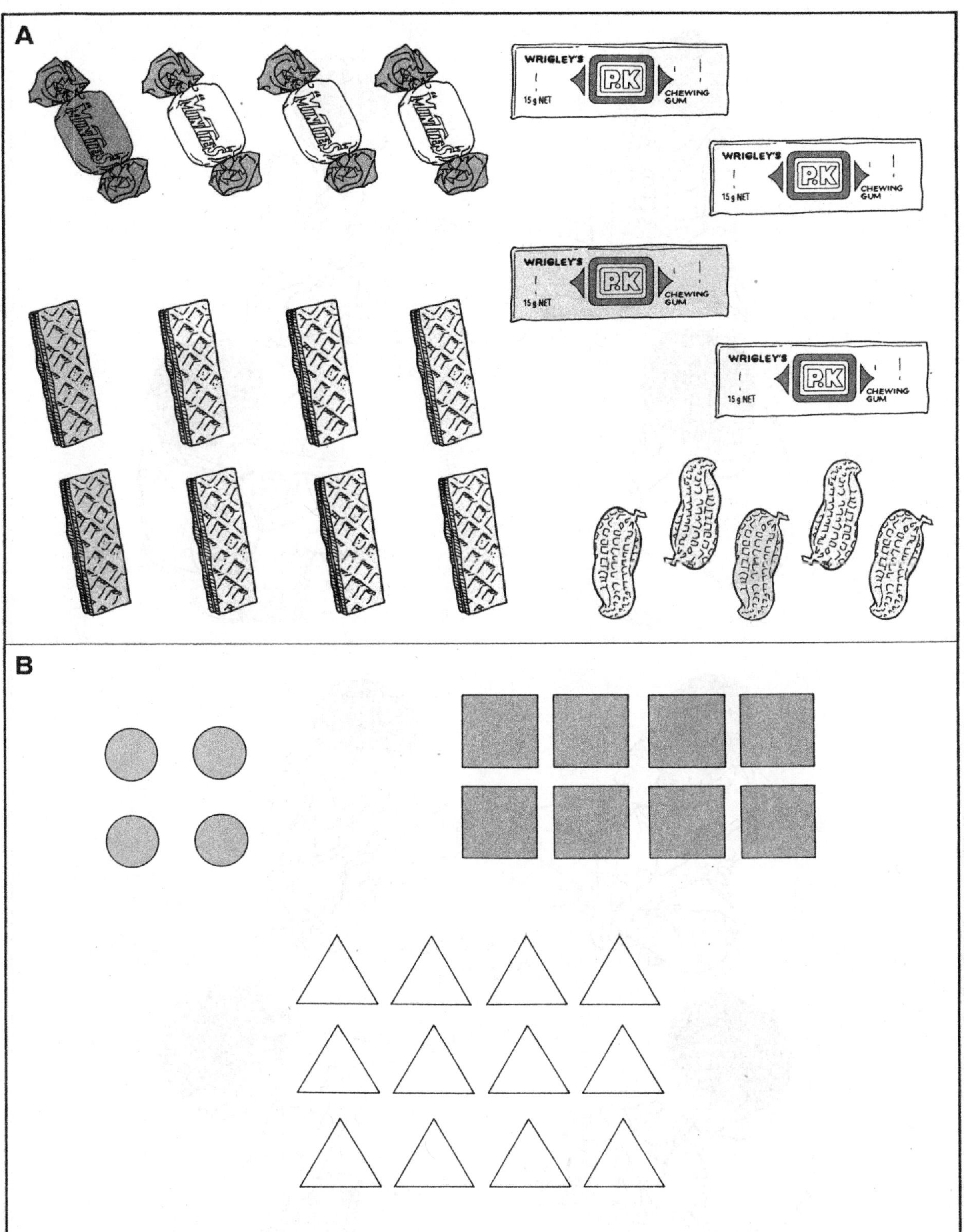
A
WRIGLEY'S
P.K
15 g NET
CHEWING GUM
B

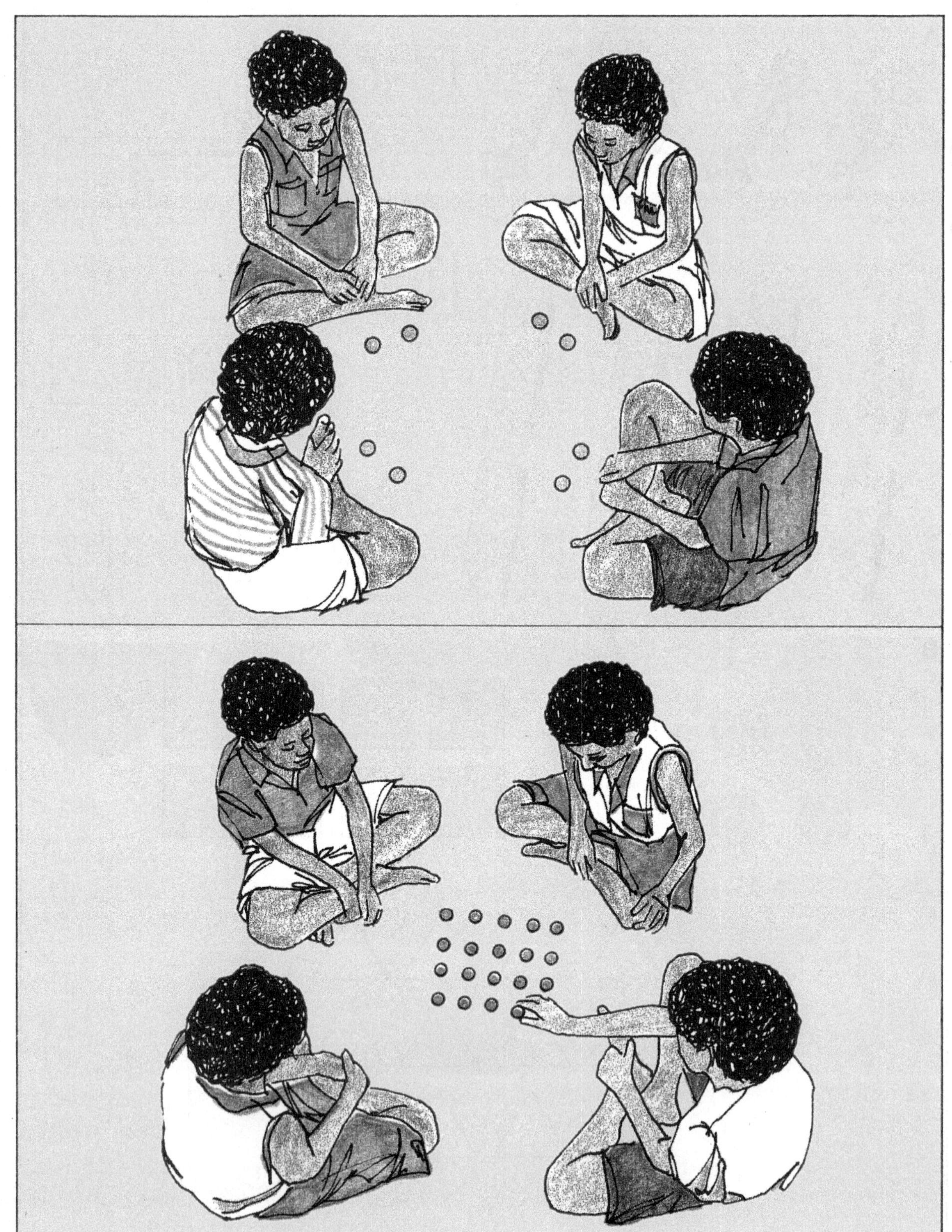

A
B
C
D

What time is it?

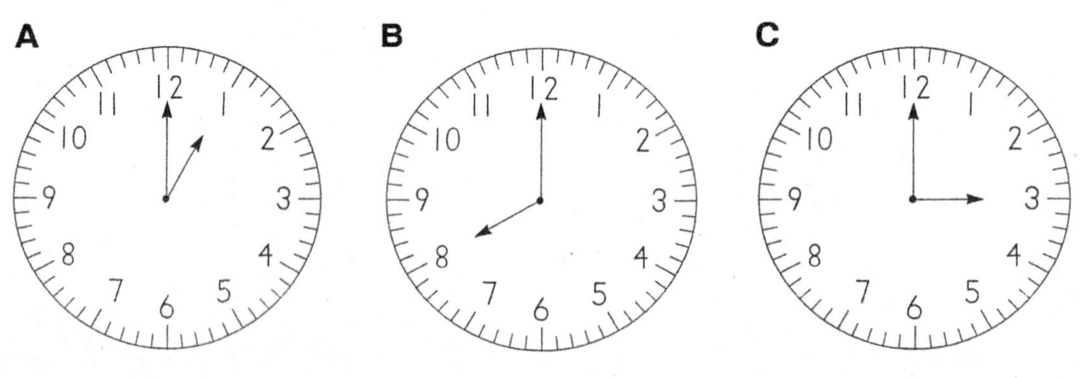
A
B
C

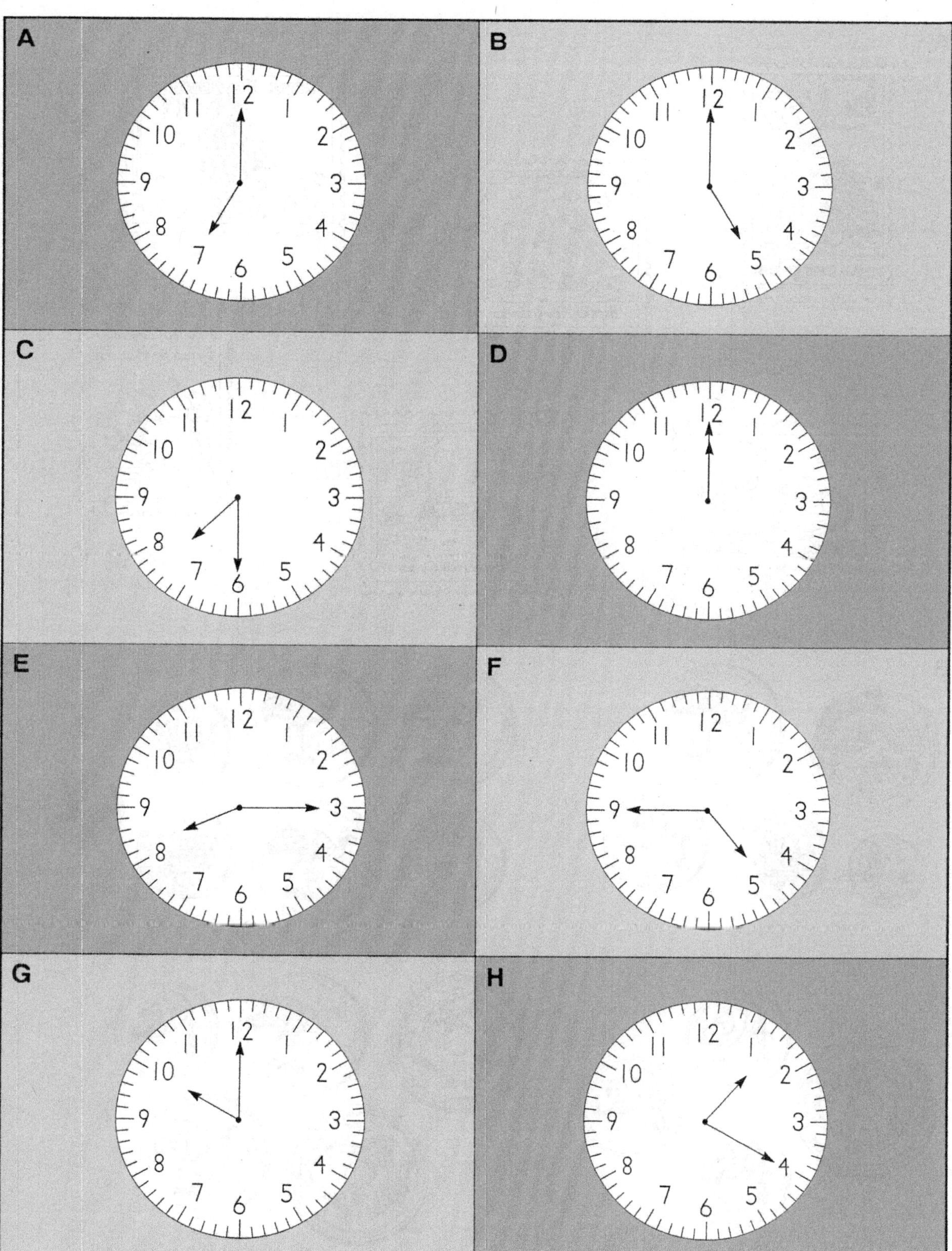
A
B
C
D
E
F
G
H

A

35t
papua new guinea
1t
Papua New Guinea
Papua New Guinea
BASILAKI MILNE BAY
70t
Papua New Guinea
21t
PAPUA NEW GUINEA
30t
Papua New Guinea
3t

B

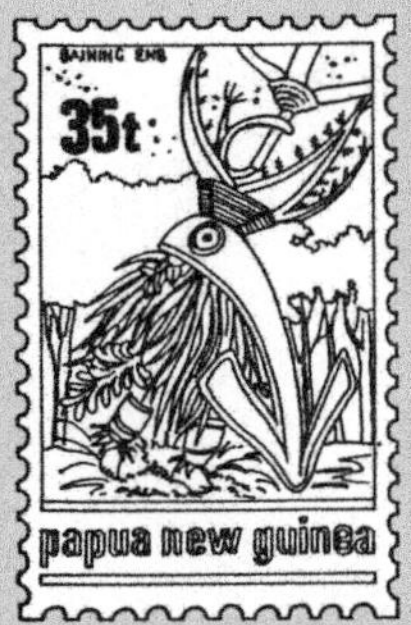

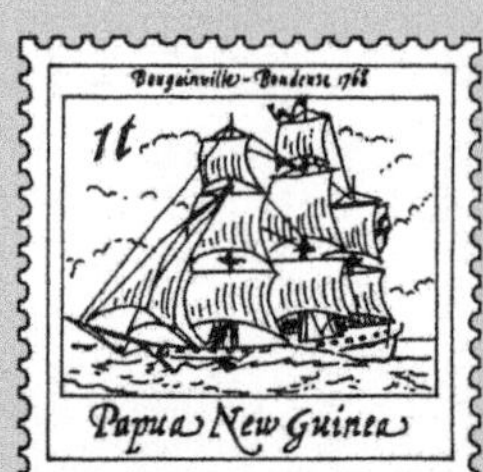

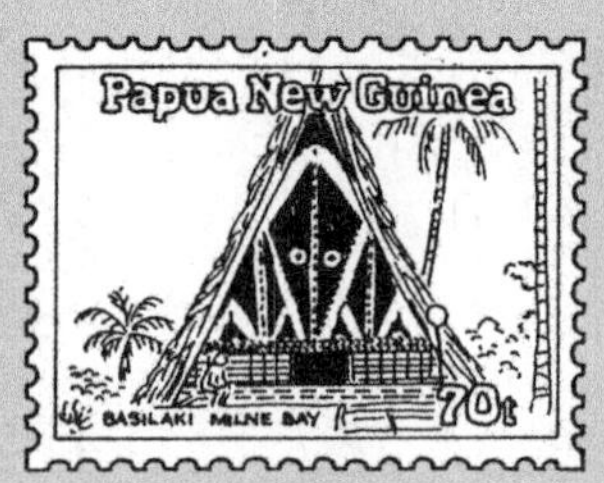

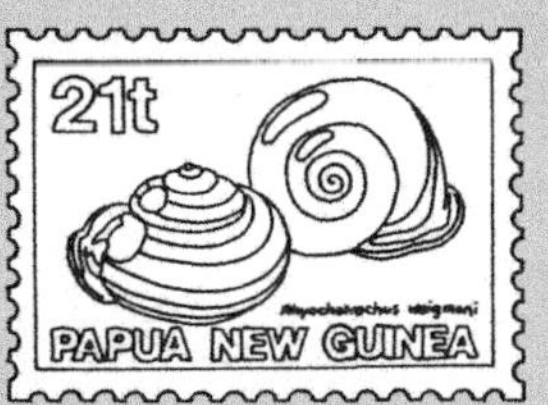

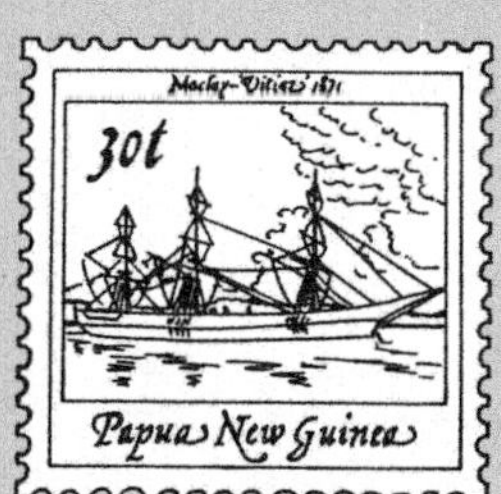

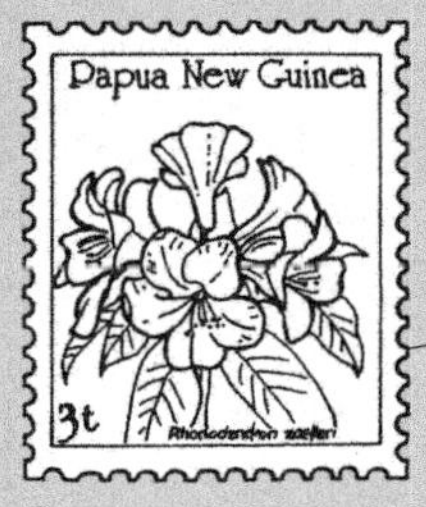

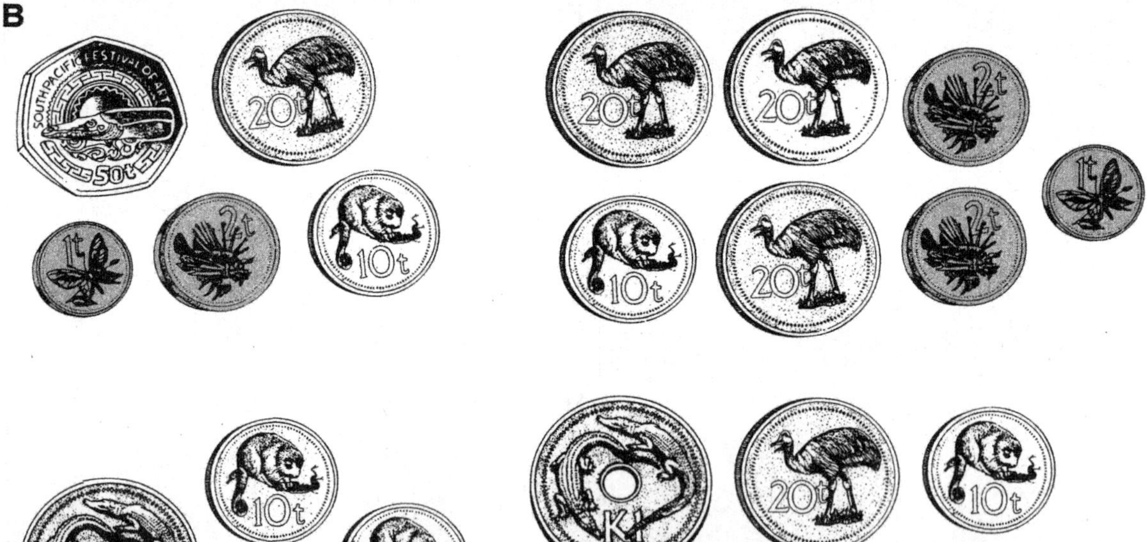
SOUTH PACIFIC FESTIVAL OF ART
50t
20t
1t
2t
10t
K1

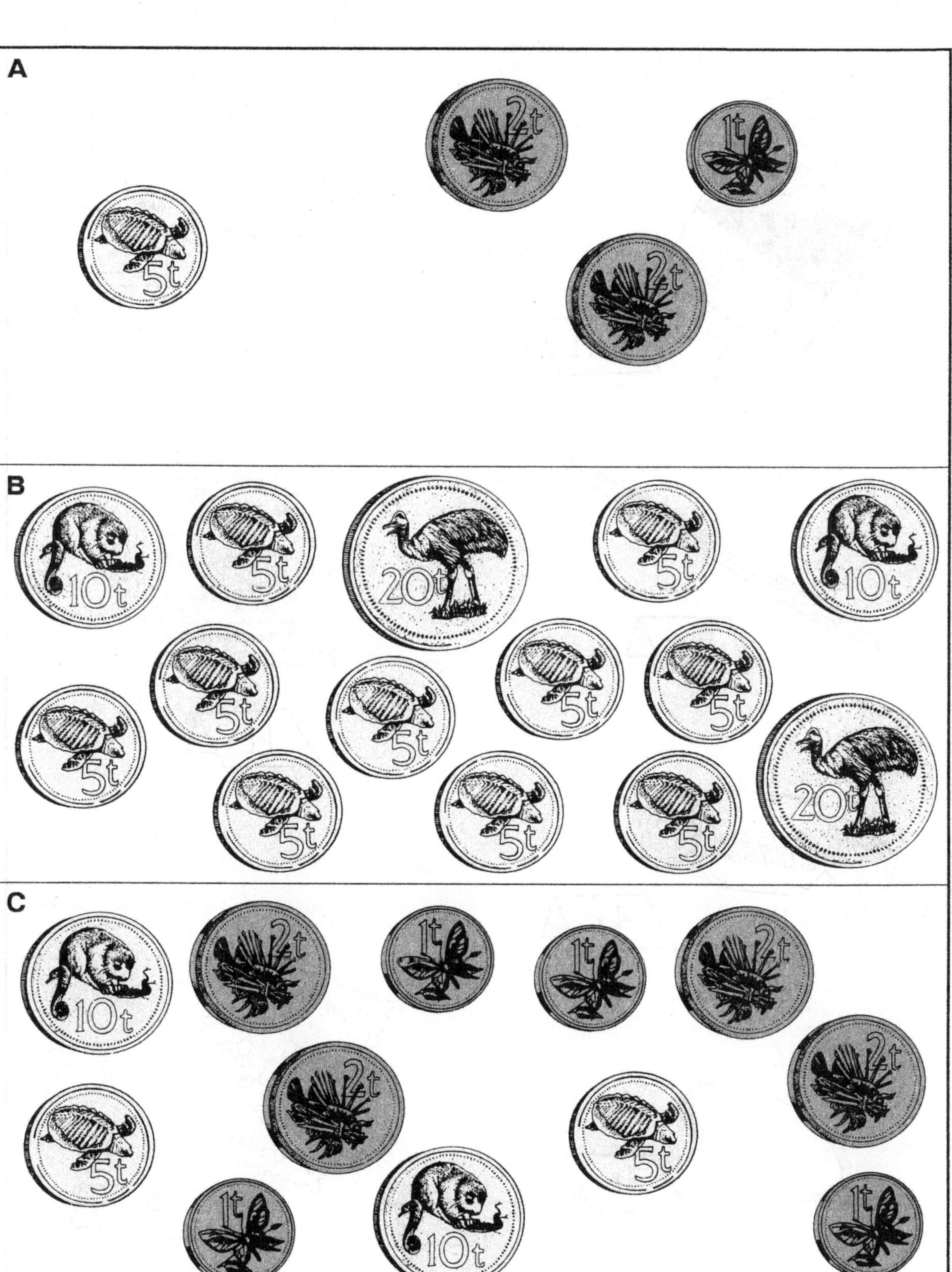
A
2t
1t
5t
2t
B
10t
5t
20t
5t
10t
5t
5t
5t
5t
5t
5t
5t
5t
20t
C
10t
2t
1t
1t
2t
2t
2t
5t
5t
1t
10t
1t

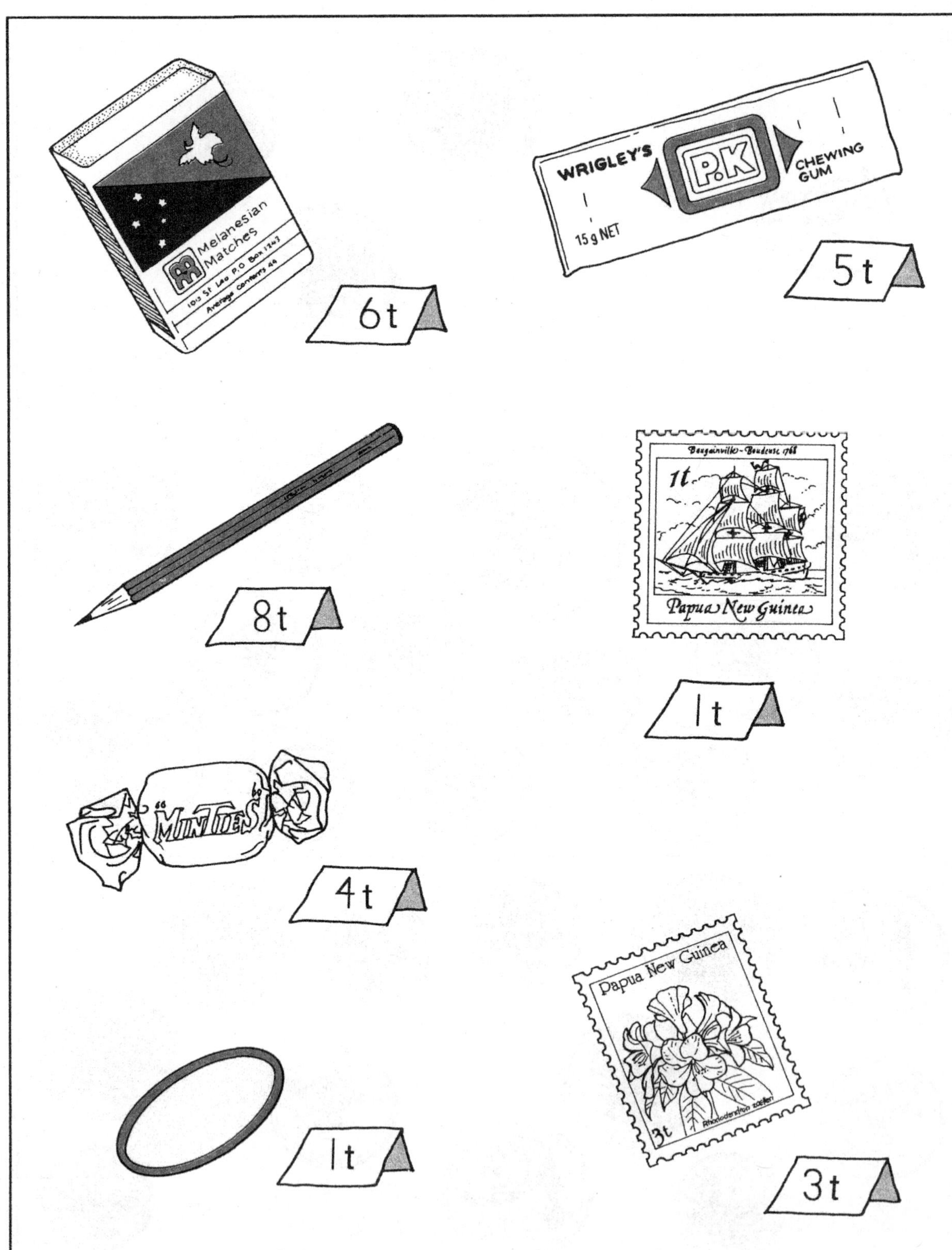
Melanesian Matches
1013 St Leo P.O Box 1243
Average Contents 44
6t
WRIGLEY'S
P.K
CHEWING GUM
15 g NET
5t
8t
Papua New Guinea
1t
1t
MINTIES
4t
1t
Papua New Guinea
3t
3t

ENRICHED
trukai
RICE
1kg NET
trukai
90t

No.1
Pineapple
Juice
DRINK
250 ml
KAPUL
Lets do it PNG
60t

PAULS
NATURAL
LONGLIFE
MILK
UHT PROCESSED
100%
PURE
250ml
PRODUCT OF AUSTRALIA

70t

560gm. Nom Mass
Arrow
Li'l Bro' Arrow says
Arrow FEIVA FLAVOUR
white sliced bread
80t

RAMU
RAMU
RAMU
SUGAR
1kg nett

70t

"777"
MACKEREL

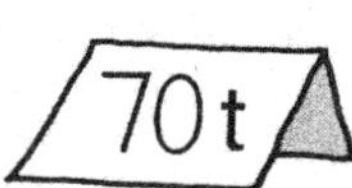

40t

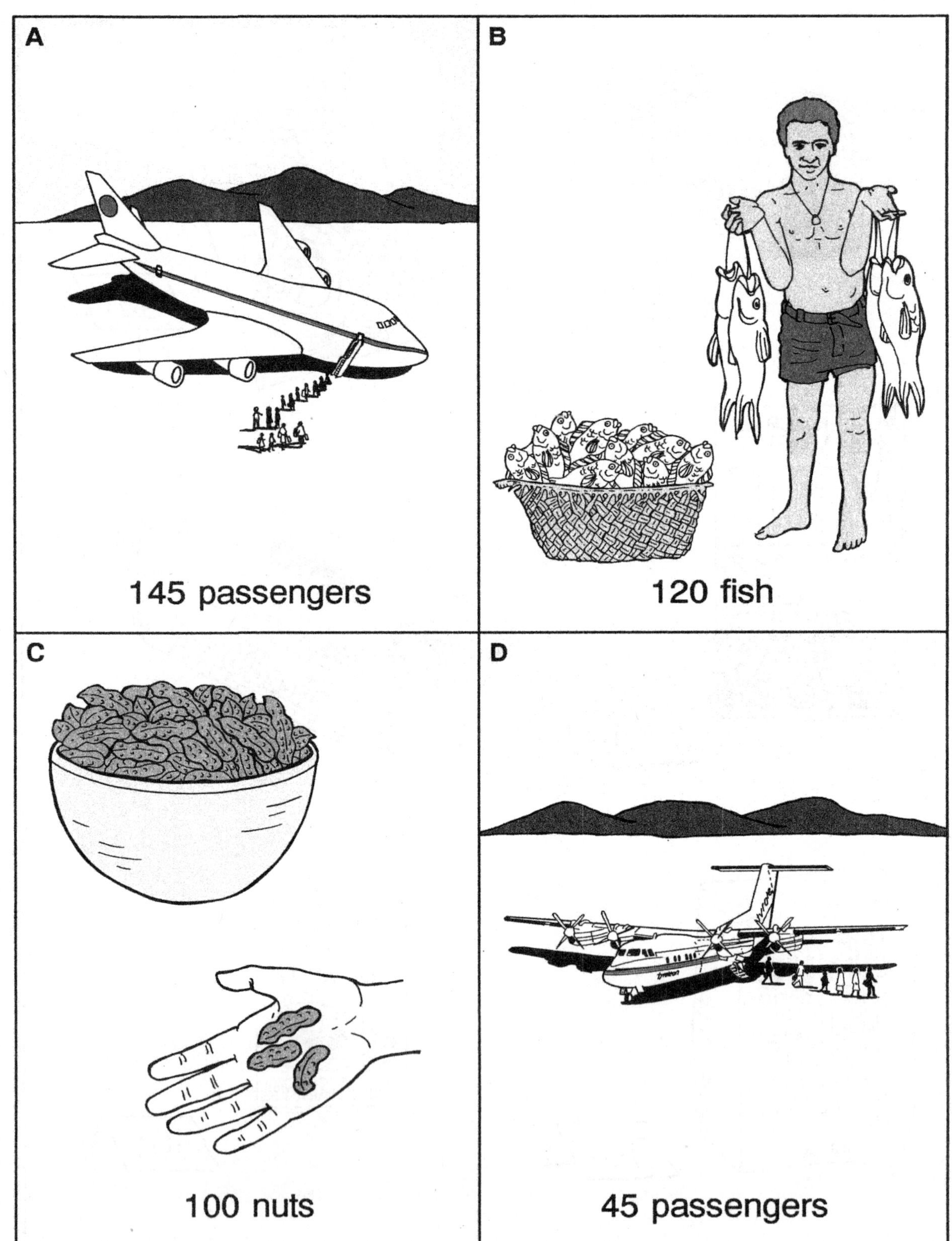
A
145 passengers
B
120 fish
C
100 nuts
D
45 passengers

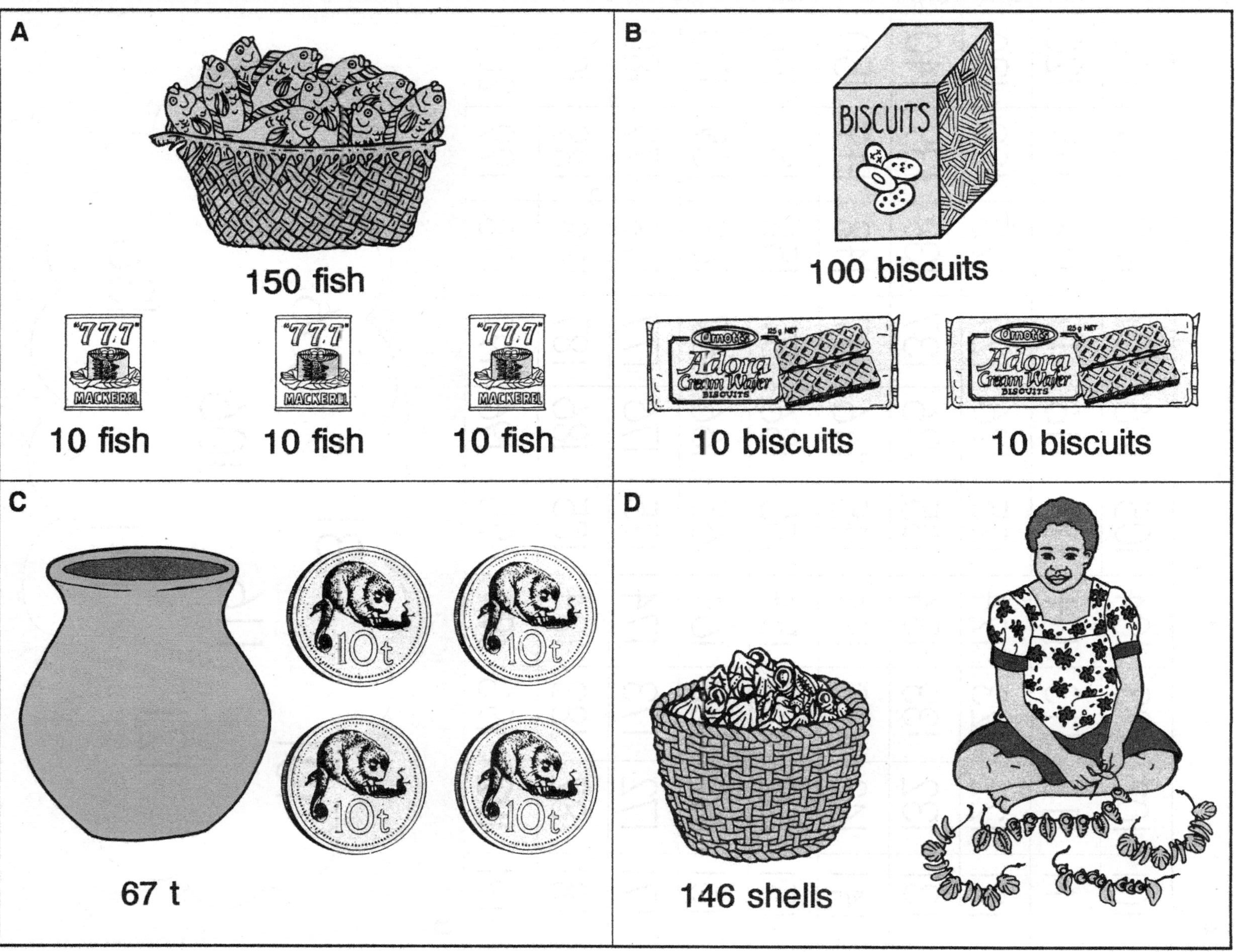
A
150 fish
"777"
MACKEREL
10 fish
10 fish
10 fish
B
BISCUITS
100 biscuits
Arnott's
Adora
Cream Wafer
BISCUITS
10 biscuits
10 biscuits
C
10t
67 t
D
146 shells

A

101	102	103	104	105	106	107	108	109	110
111	112	113	114	115	116	117	118	119	120
121	122	123	124	125	126	127	128	129	130
131	132	133	134	135	136	137	138	139	140
141	142	143	144	145	146	147	148	149	150
151	152	153	154	155	156	157	158	159	160
161	162	163	164	165	166	167	168	169	170
171	172	173	174	175	176	177	178	179	180
181	182	183	184	185	186	187	188	189	190
191	192	193	194	195	196	197	198	199	200

B

131 156

101

116 106

121

141 136

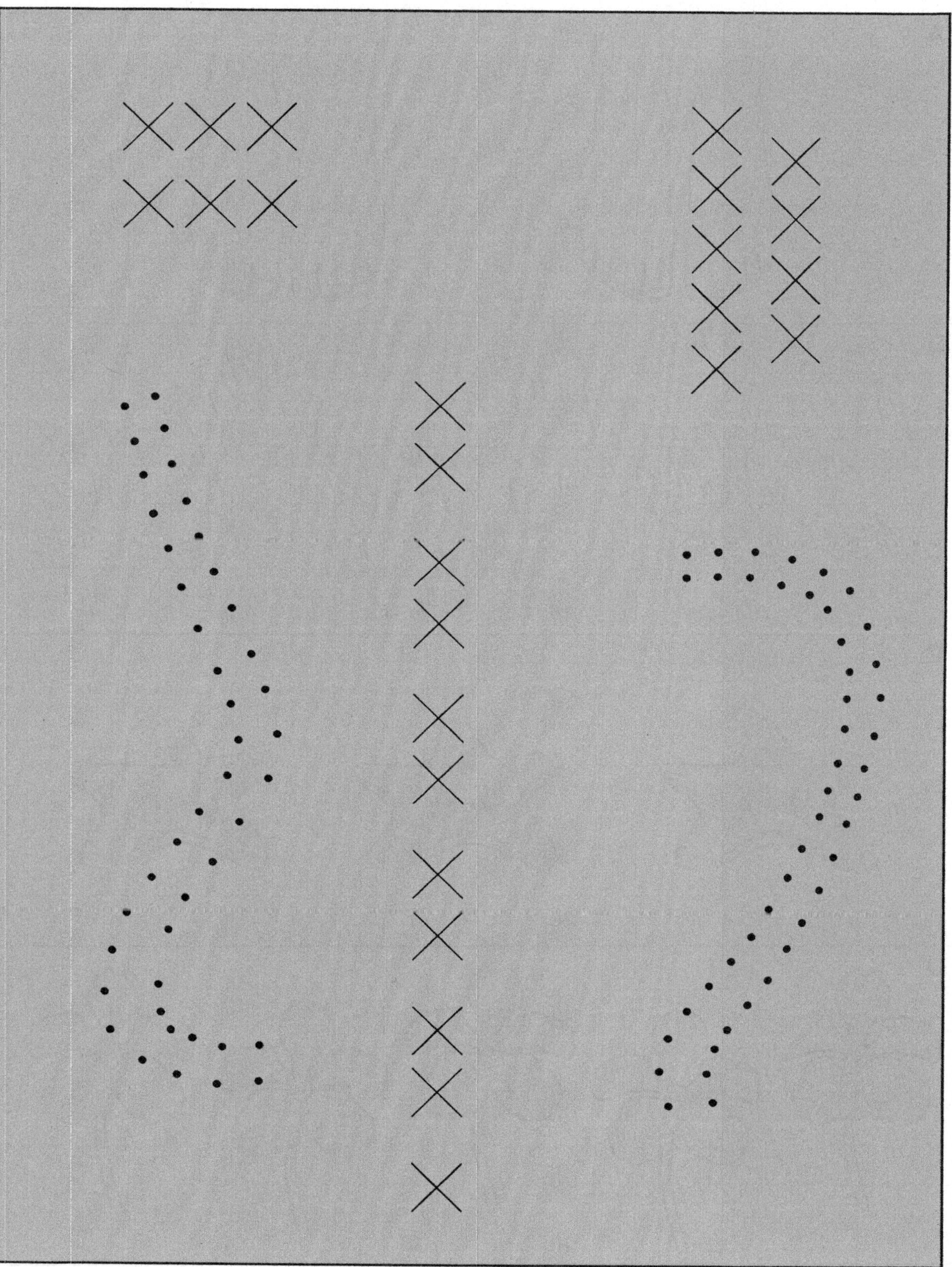

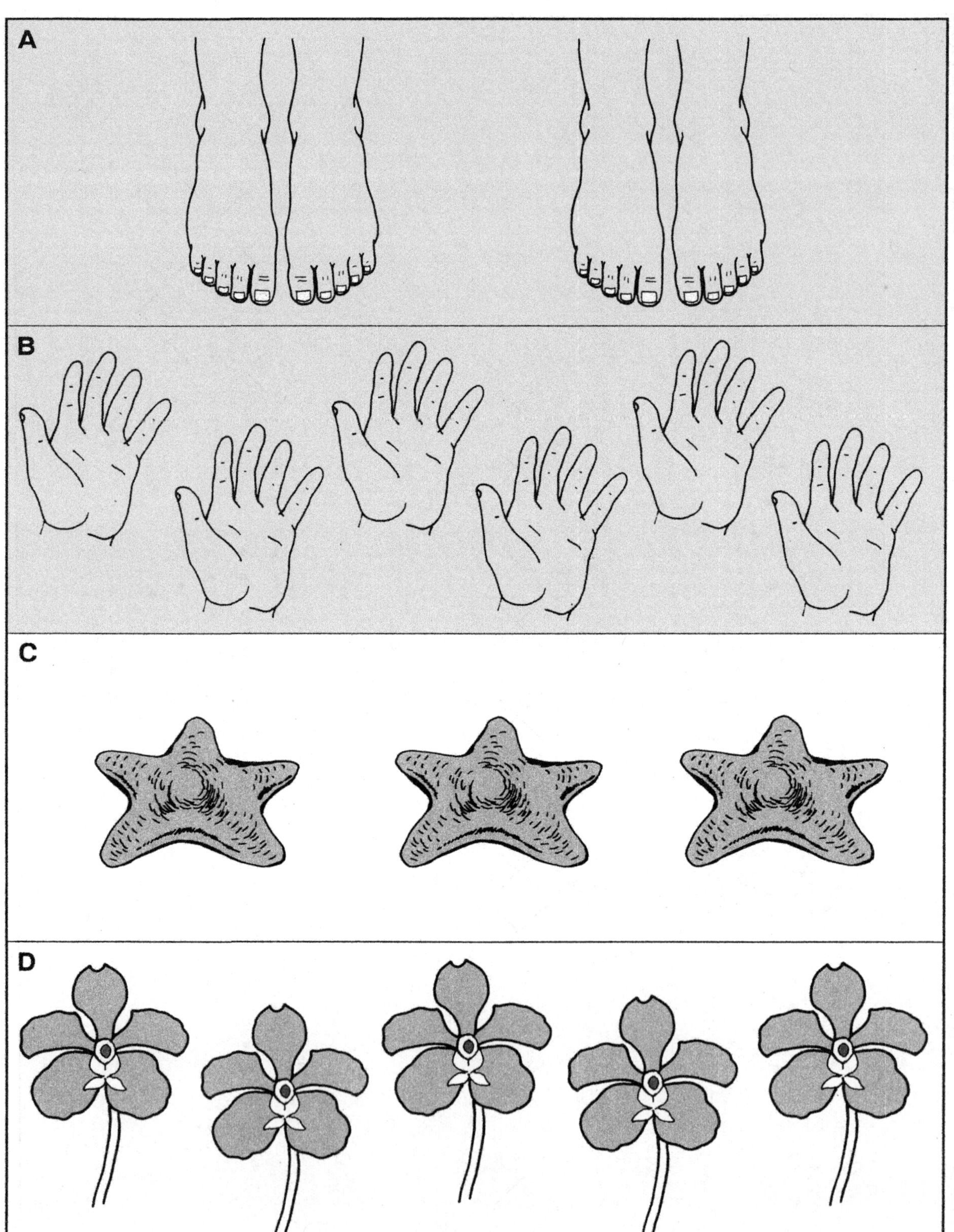
A
B
C
D

A

B

C

D

E

F

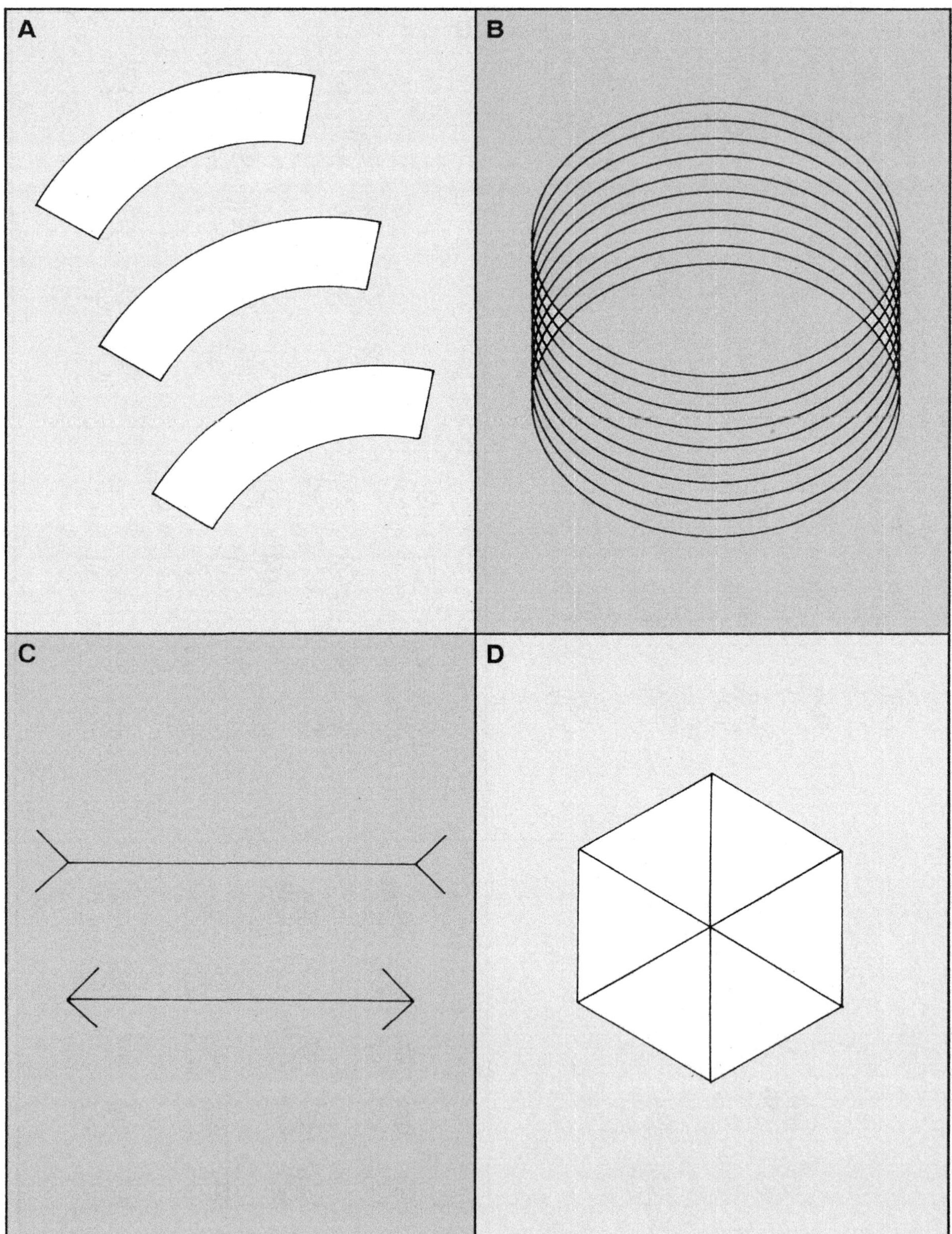
A
B
C
D

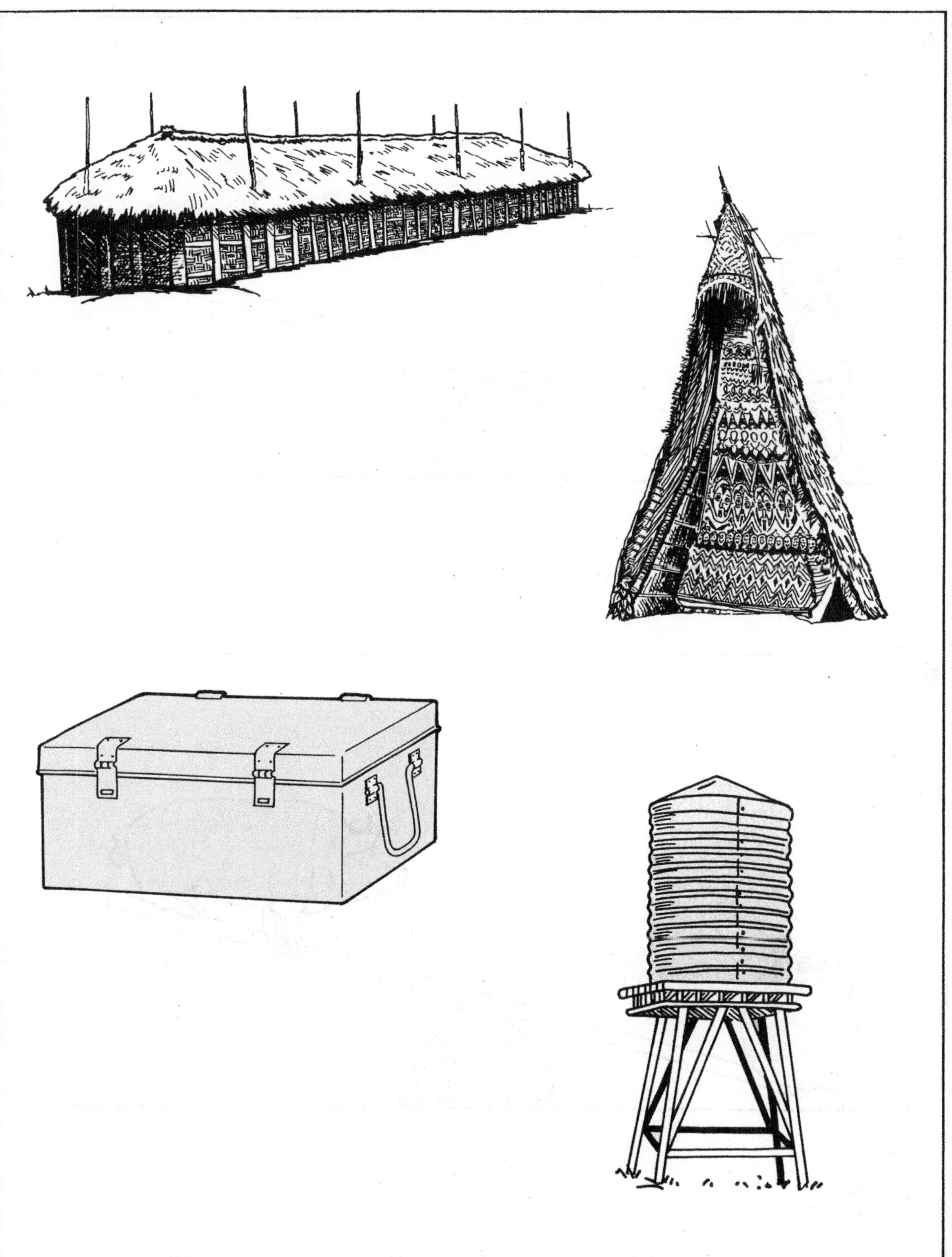

A
B

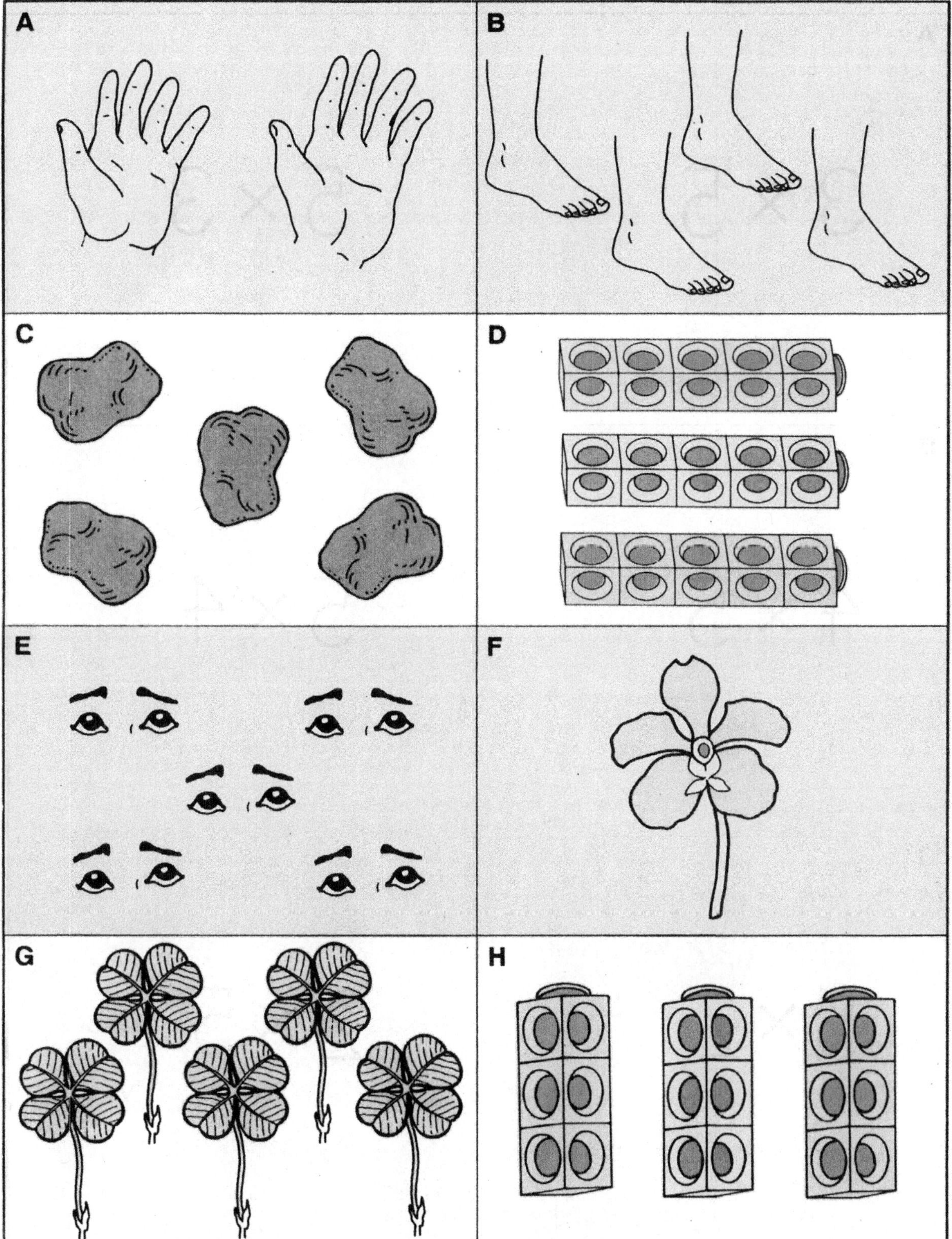
A
B
C
D
E
F
G
H

A	
3×5	5×3

B	
4×5	5×4

C	
5×2	2×5

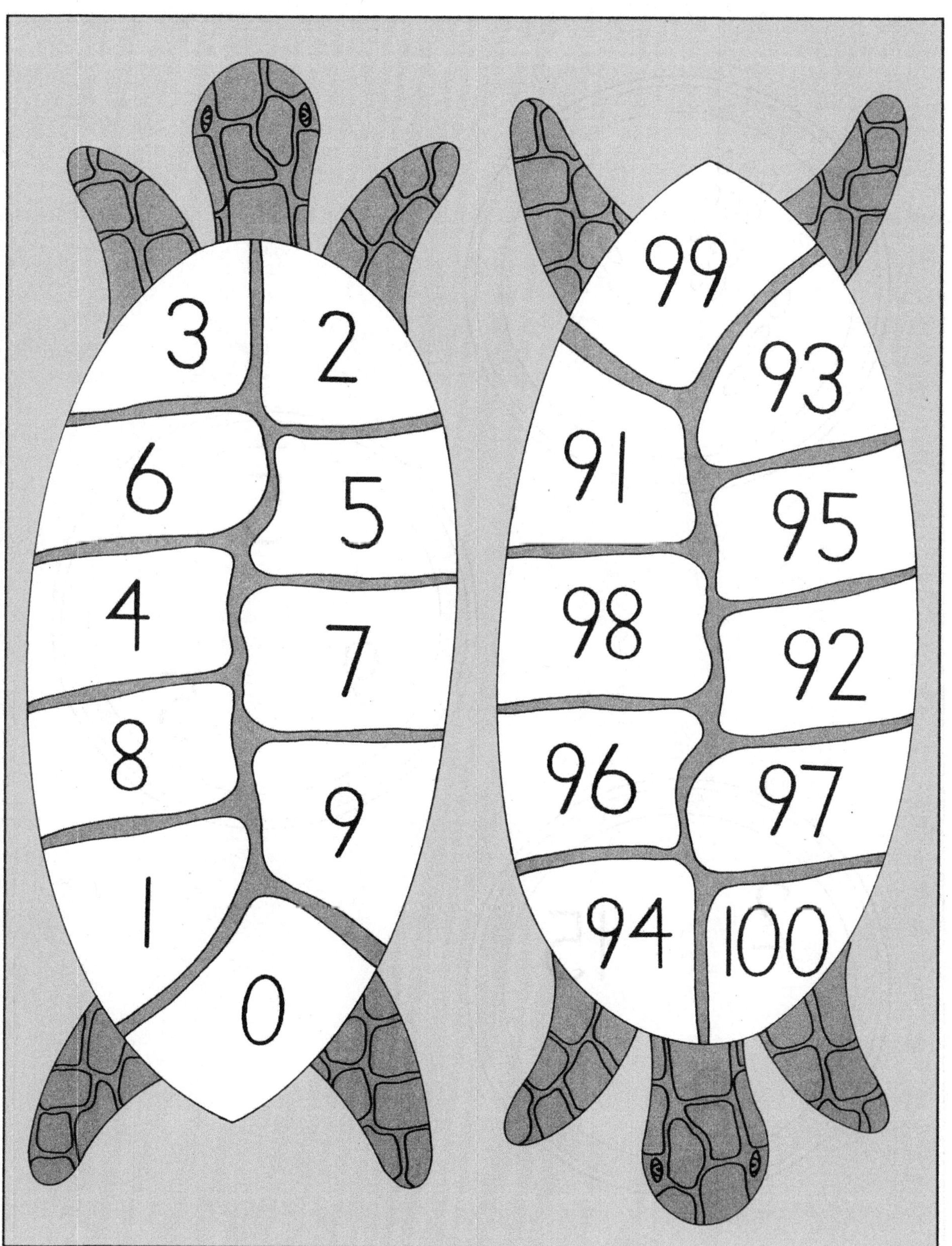
3
2
6
5
4
7
8
9
1
0
99
93
91
95
98
92
96
97
94
100

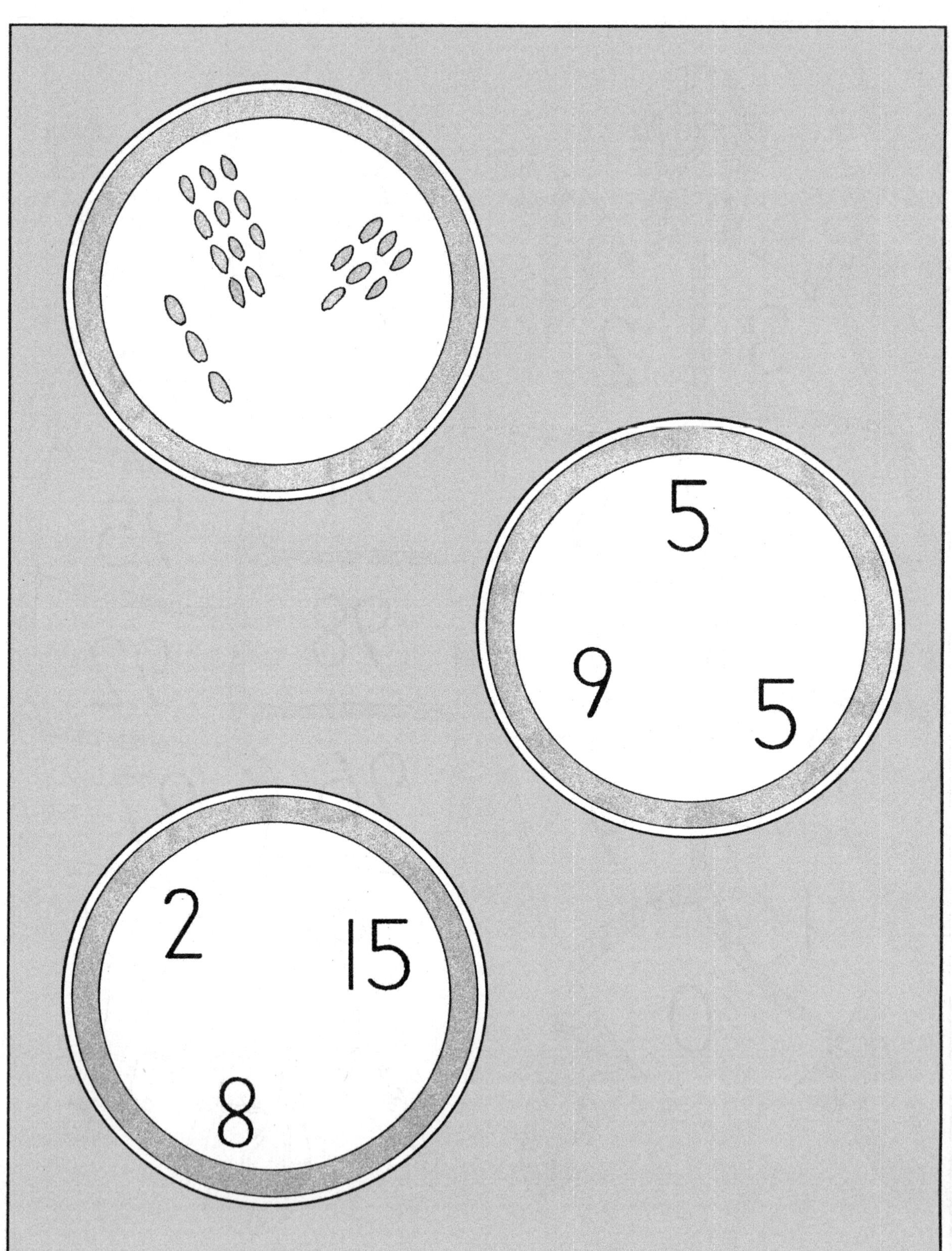
5
9
5
2
15
8

$3+3=$ $10+11=$

$5+5=$ $4+4=$

$20+20=$ $4+5=$

$5+6=$ $2+3=$

$7+7=$ $6+6=$

$3+4=$ $6+7=$

15 + 10 =

22 + 10 =

32 + 10 =

10 + 22 =

131 + 10 =

151 + 10 =

22 + 9 =

32 + 9 =

15 + 9 =

131 + 9 =

151 + 9 =

172 + 9 =

$7 + 20 =$

$31 + 20 =$

$20 + 20 =$

$20 + 15 =$

$20 + 25 =$

$37 + 20 =$

$141 + 20 =$

$110 + 20 =$

$140 + 20 =$

$20 + 150 =$

$180 + 20 =$

$17 + 20 =$

110	111	112	113	114	115	116	117	118	119
120	121	122	123	124	125	126	127	128	129
130	131	132	133	134	135	136	137	138	139
140	141	142	143	144	145	146	147	148	149
150	151	152	153	154	155	156	157	158	159
160	161	162	163	164	165	166	167	168	169
170	171	172	173	174	175	176	177	178	179
180	181	182	183	184	185	186	187	188	189
190	191	192	193	194	195	196	197	198	199
200	201	202	203	204	205	206	207	208	209
210	211	212	213	214	215	216	217	218	219
220	221	222	223	224	225	226	227	228	229
230	231	232	233	234	235	236	237	238	239
240	241	242	243	244	245	246	247	248	249
250	251	252	253	254	255	256	257	258	259
260	261	262	263	264	265	266	267	268	269
270	271	272	273	274	275	276	277	278	279
280	281	282	283	284	285	286	287	288	289
290	291	292	293	294	295	296	297	298	299
300	301	302	303	304	305	306	307	308	309

90	91	92	93	94	95	96	97	98	99
100	101	102	103	104	105	106	107	108	109
110	111	112	113	114	115	116	117	118	119
120	121	122	123	124	125	126	127	128	129
130	131	132	133	134	135	136	137	138	139
140	141	142	143	144	145	146	147	148	149
150	151	152	153	154	155	156	157	158	159
160	161	162	163	164	165	166	167	168	169
170	171	172	173	174	175	176	177	178	179
180	181	182	183	184	185	186	187	188	189
190	191	192	193	194	195	196	197	198	199

$30 - 20 =$

$45 - 20 =$

$57 - 20 =$

$92 - 20 =$

$141 - 20 =$

$32 - 20 =$

$170 - 20 =$

$140 - 20 =$

$89 - 20 =$

$76 - 20 =$

$77 - 20 =$

$61 - 20 =$

6
7
5
6
8
3
9
7
7
9
7
6
6
6

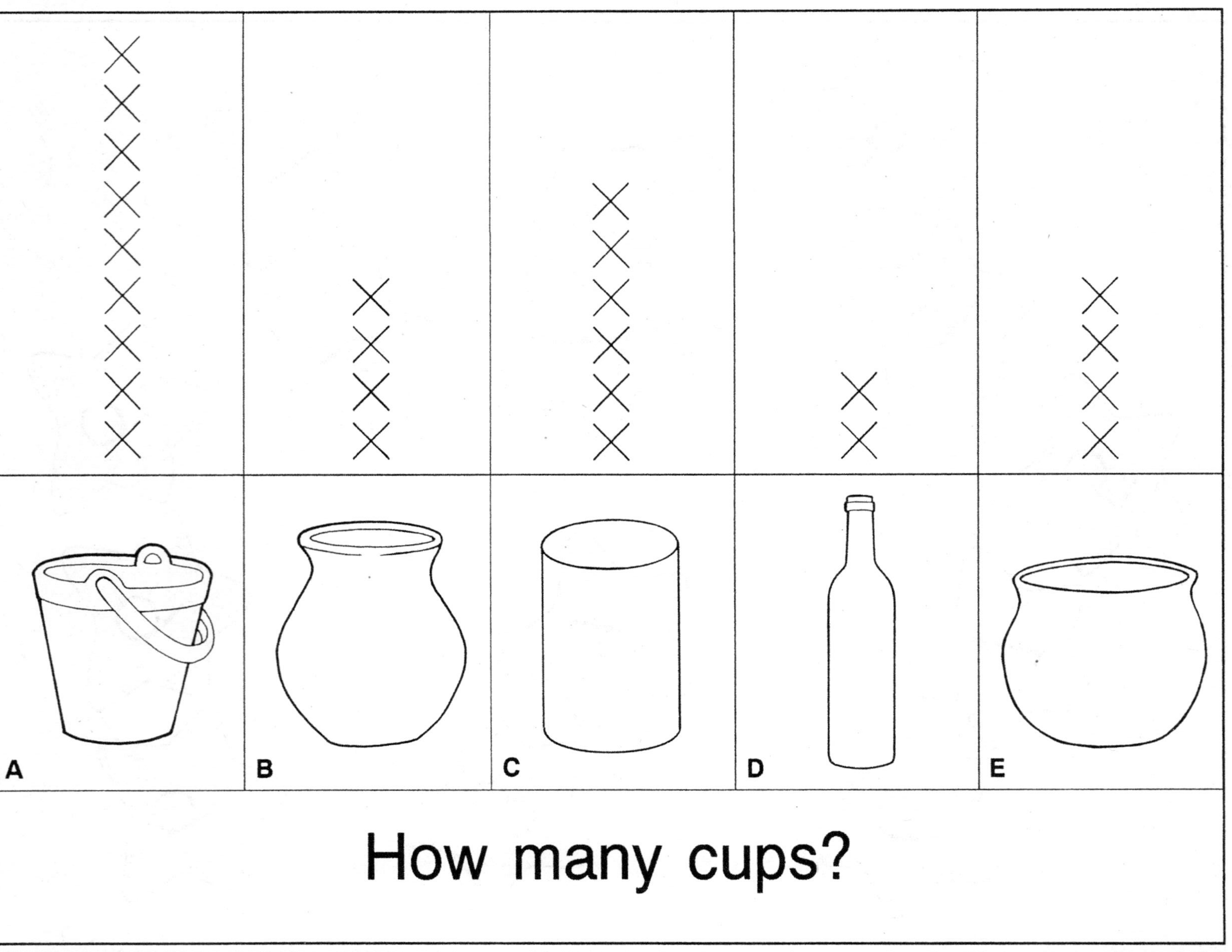
A
B
C
D
E
How many cups?

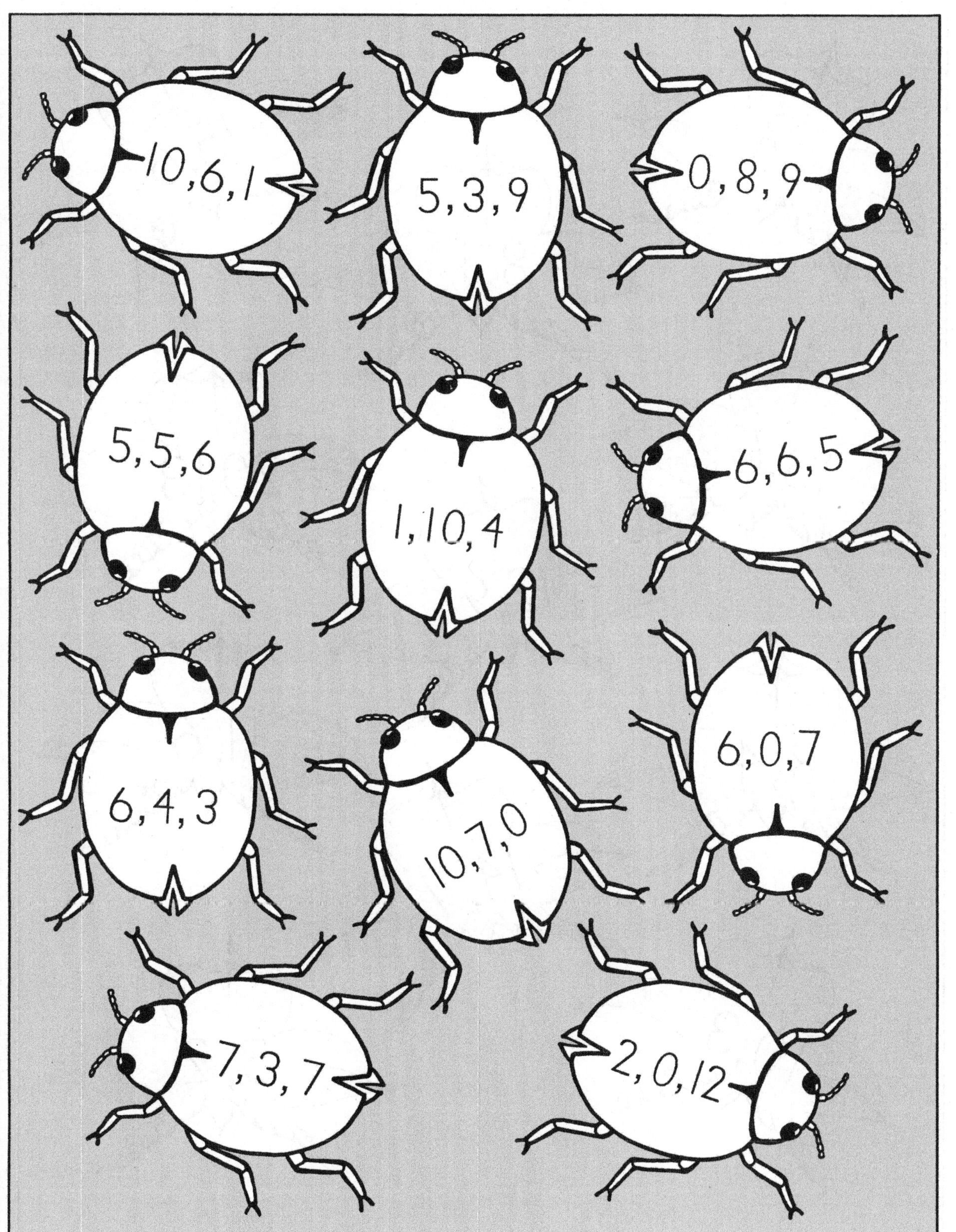
10,6,1
5,3,9
0,8,9
5,5,6
1,10,4
6,6,5
6,4,3
10,7,0
6,0,7
7,3,7
2,0,12

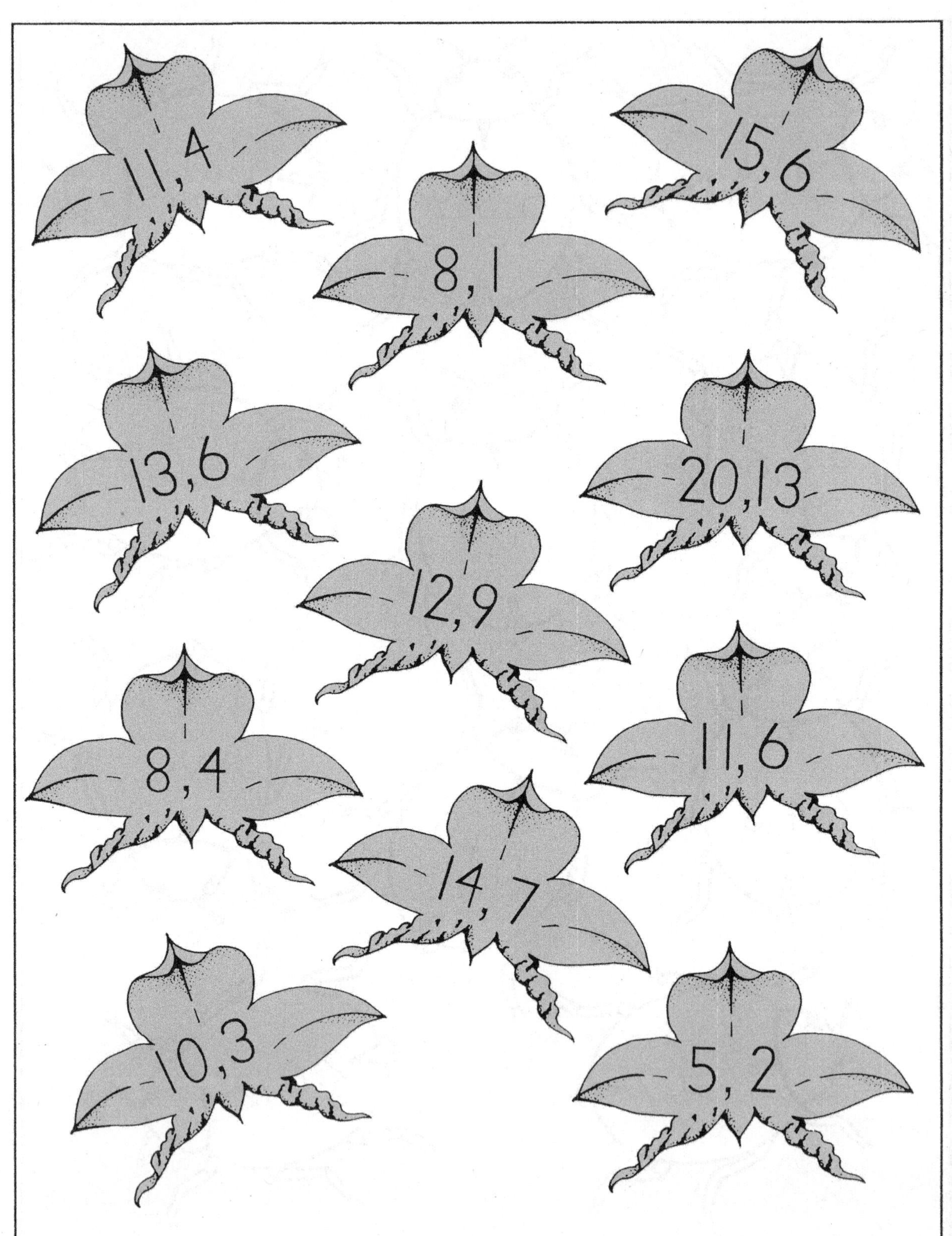
11,4
8,1
15,6
13,6
20,13
12,9
8,4
11,6
14,7
10,3
5,2

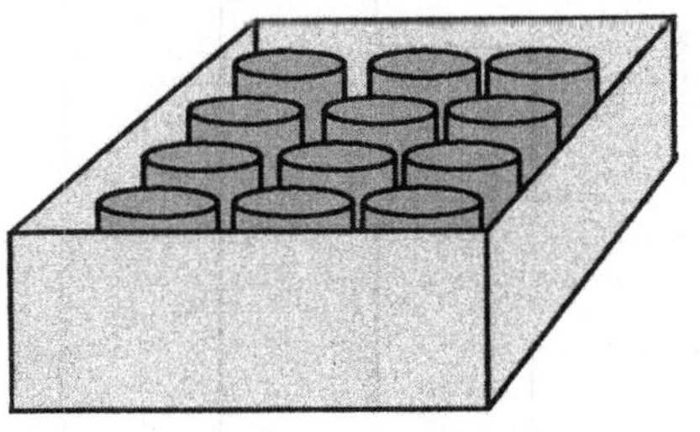

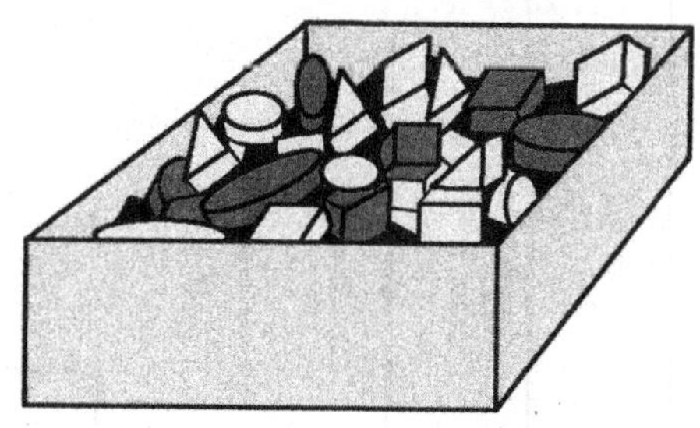

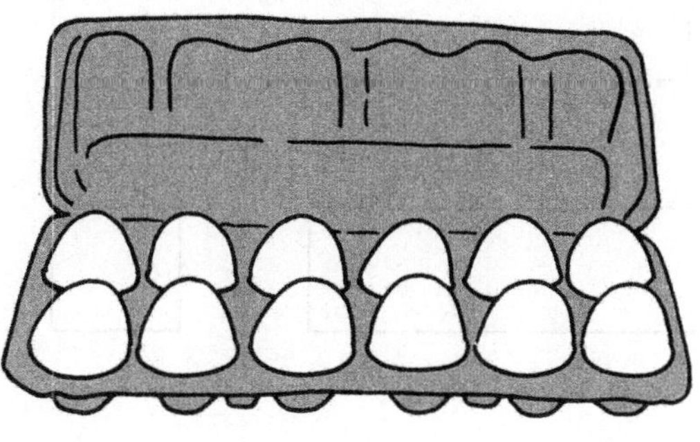

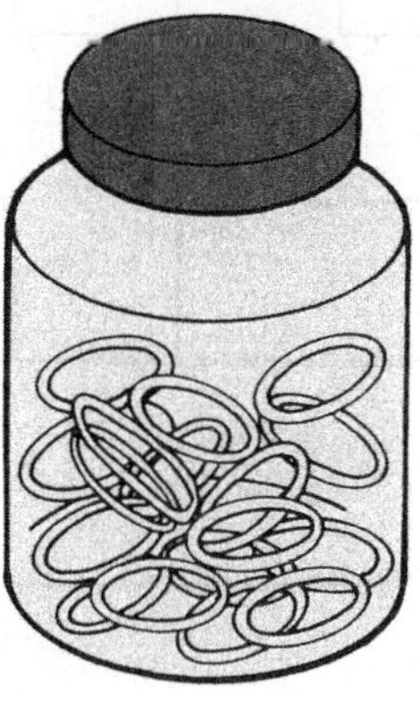

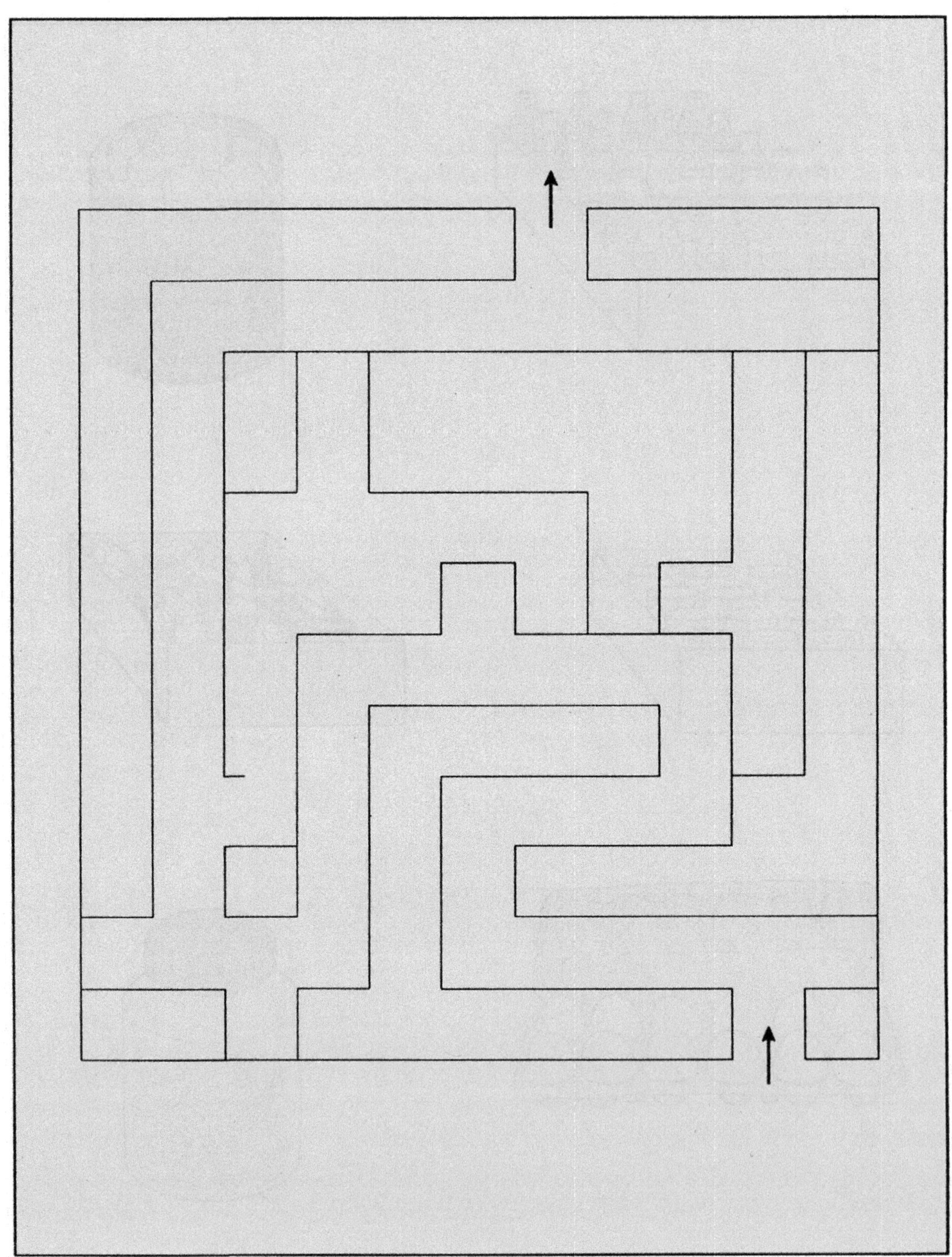

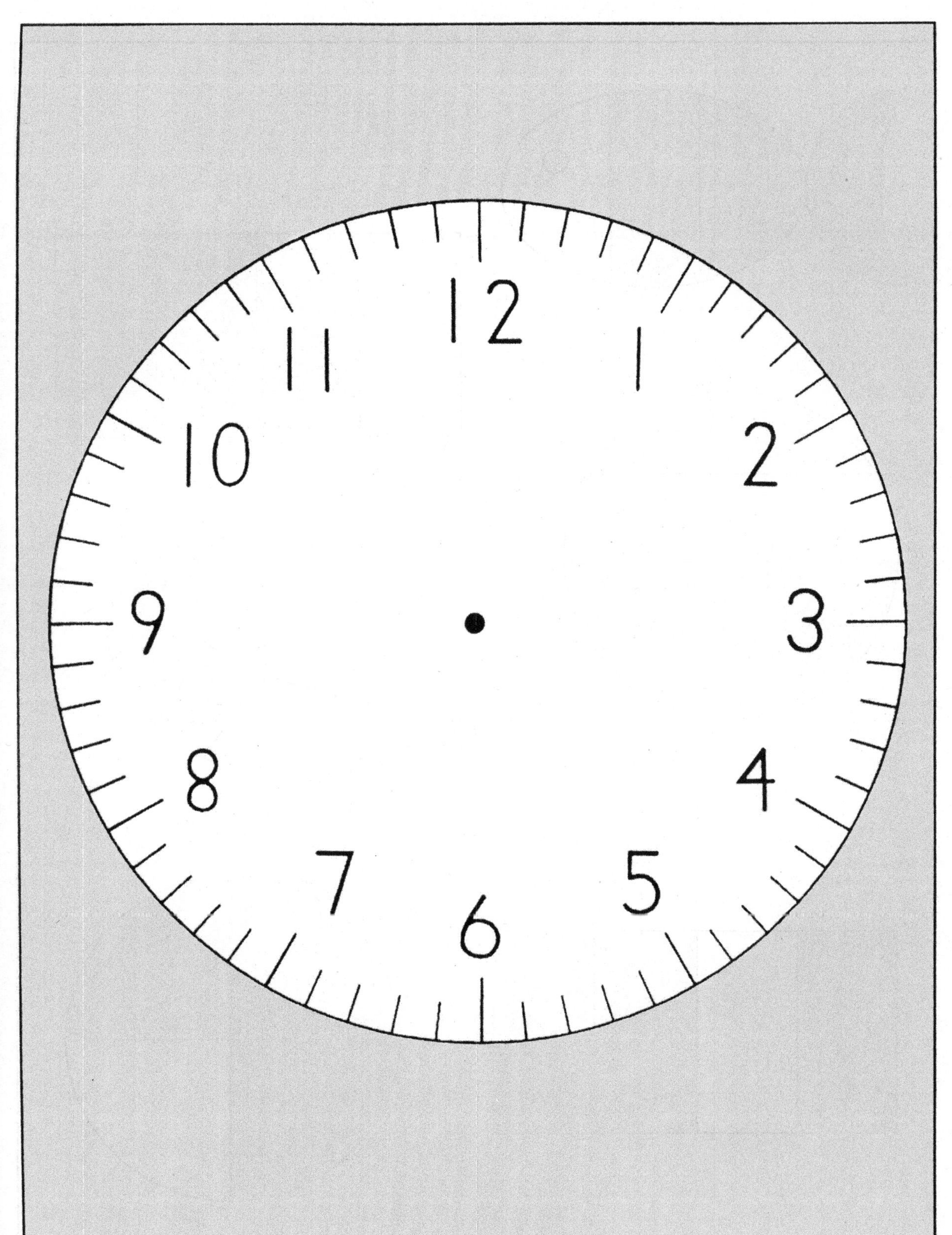
12
1
2
3
4
5
6
7
8
9
10
11

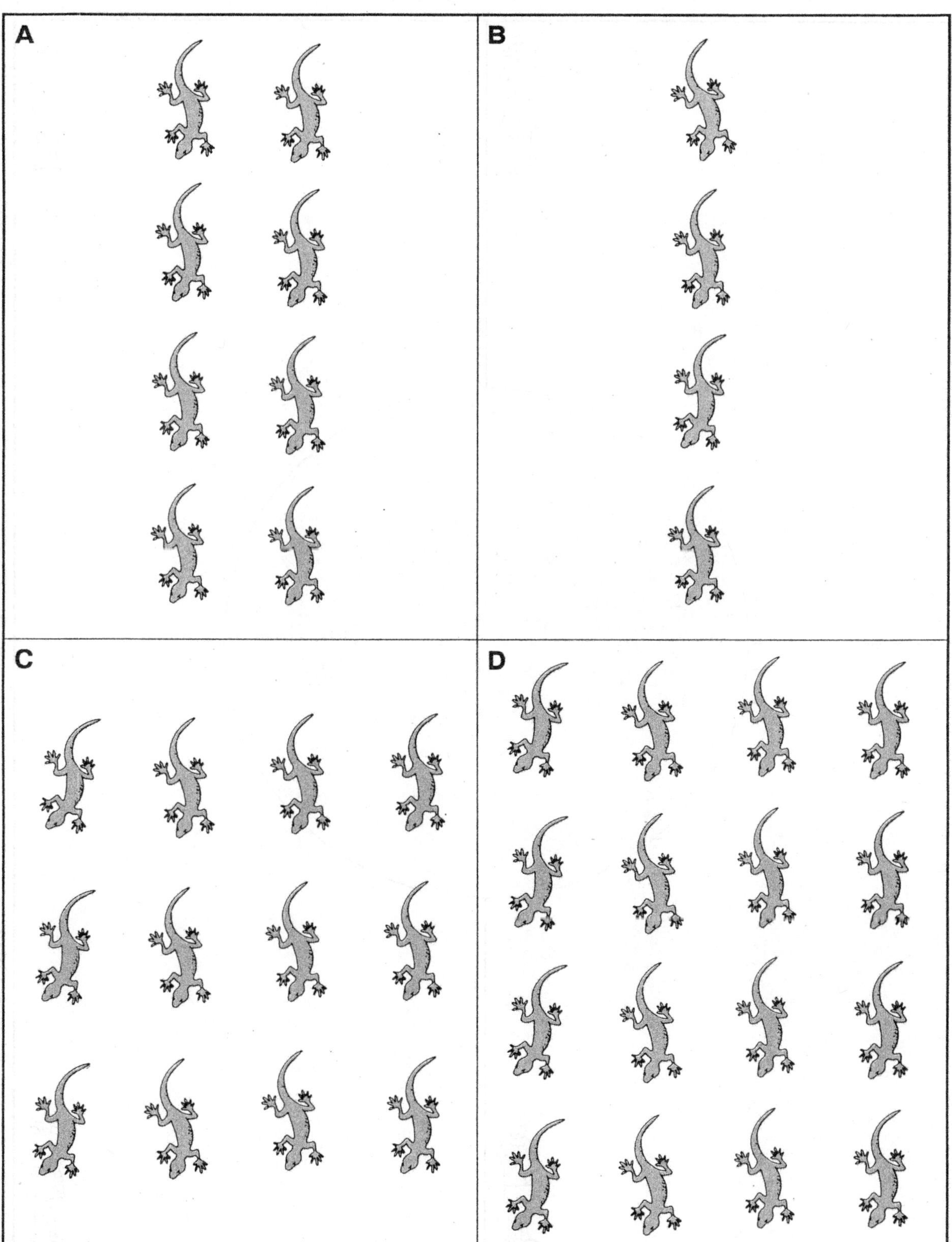
A
B
C
D

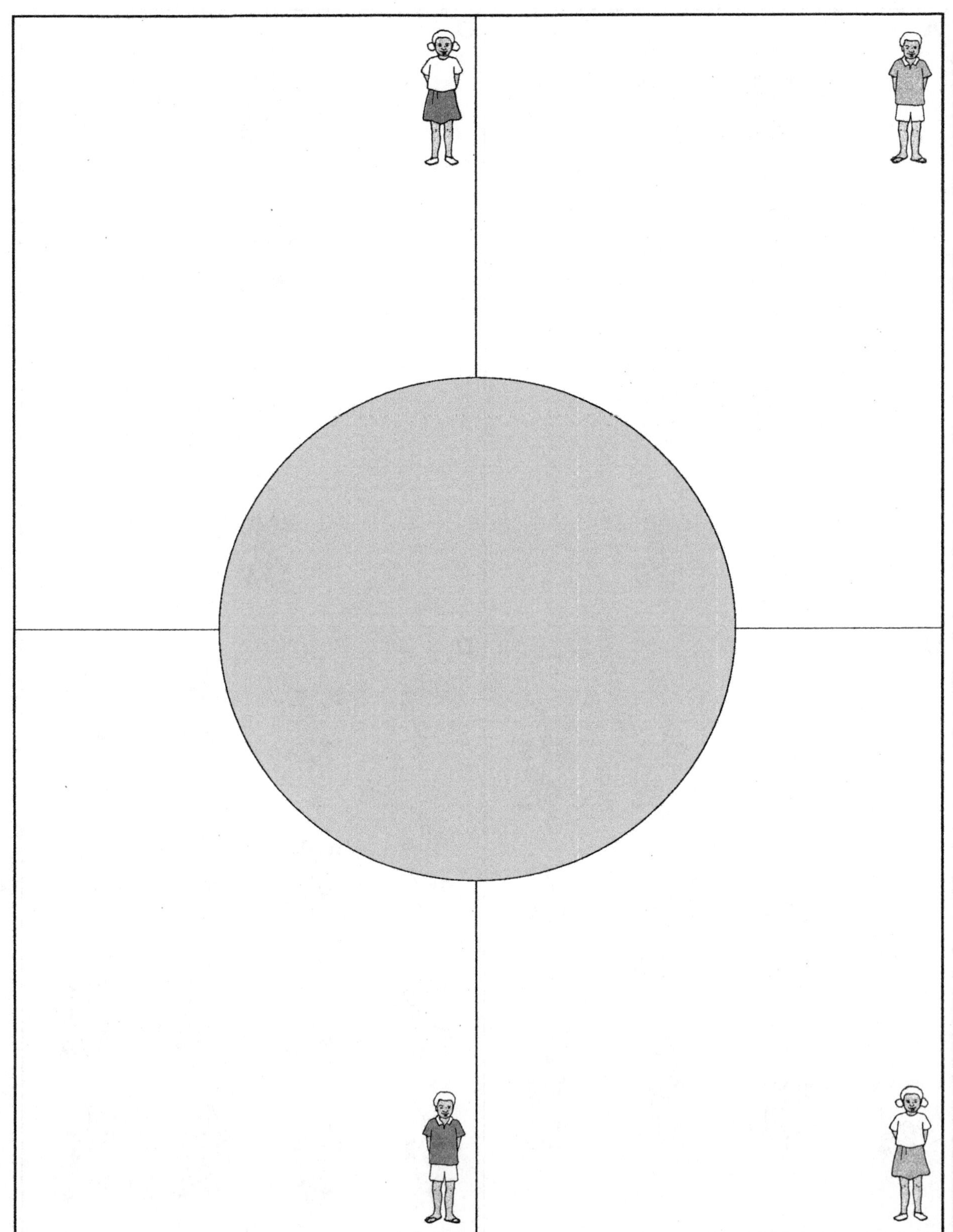

A

Christmas

B

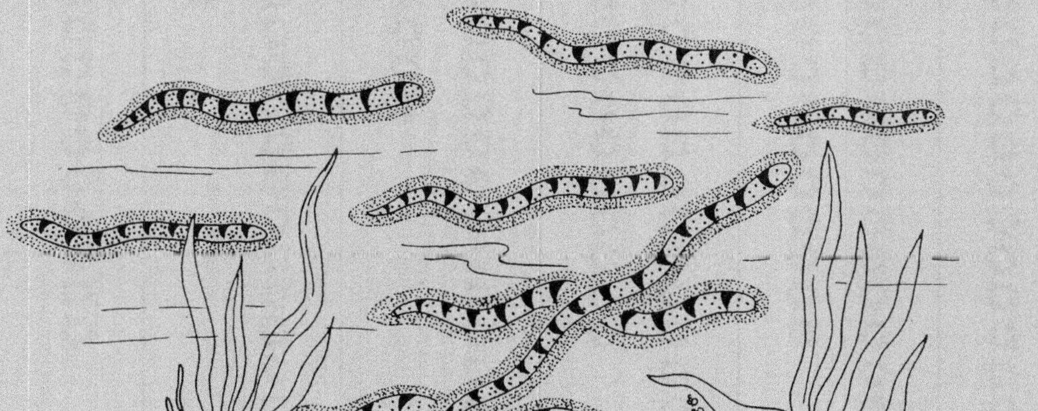

The New Year

January	February	March	April	May	June

July	August	September	October	November	December

How many sunny days in each month?

6 + 5 =	7 + 6 =	5 + 8 =	8 + 3 =	8 + 7 =	9 + 5 =
4 + 9 =	6 + 6 =	8 + 9 =	10 + 7 =	9 + 3 =	8 + 4 =
6 + 8 =	7 + 9 =	3 + 10 =	10 + 5 =	4 + 7 =	7 + 7 =
8 + 8 =	9 + 10 =	10 + 1 =	10 + 8 =	7 + 4 =	6 + 10 =

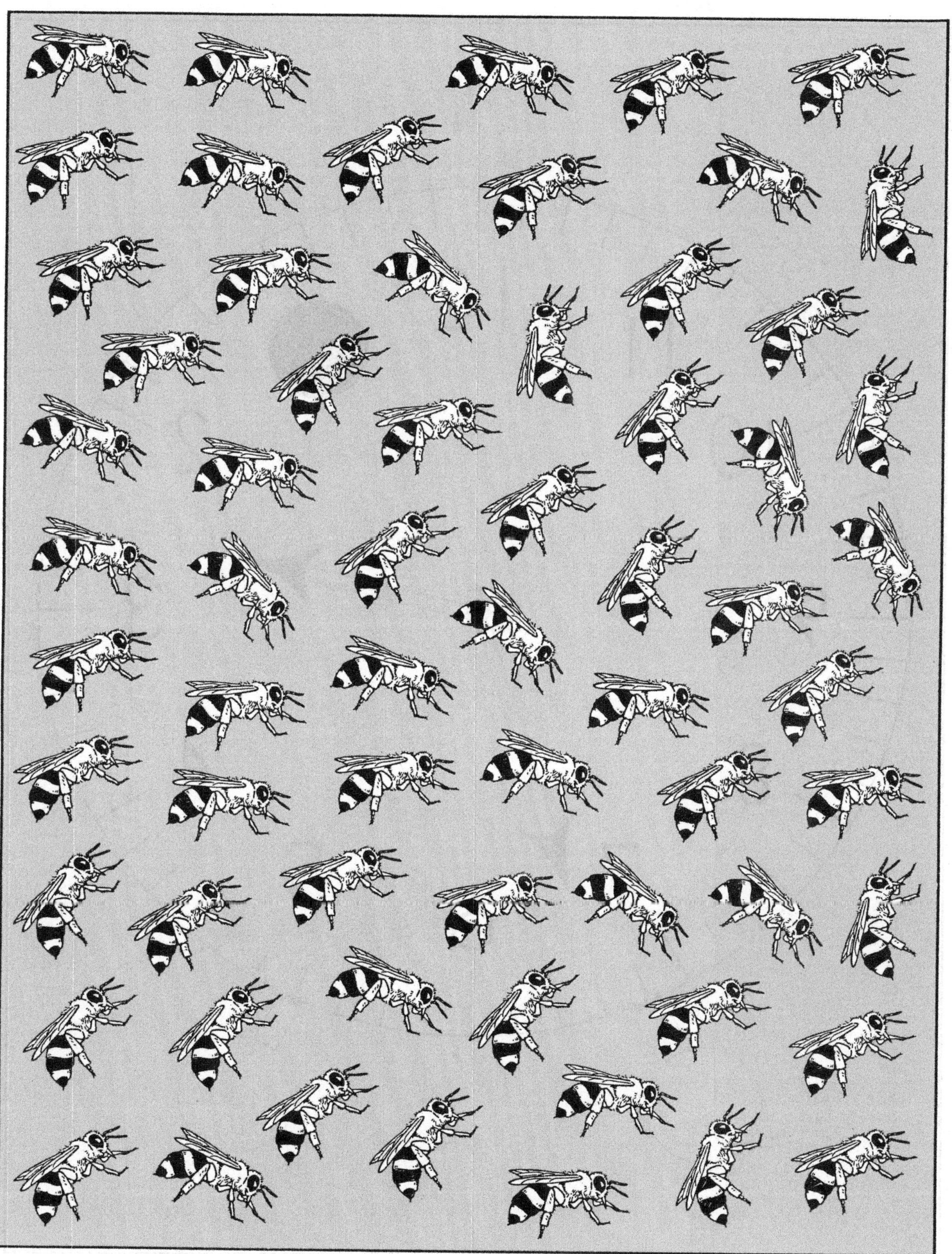

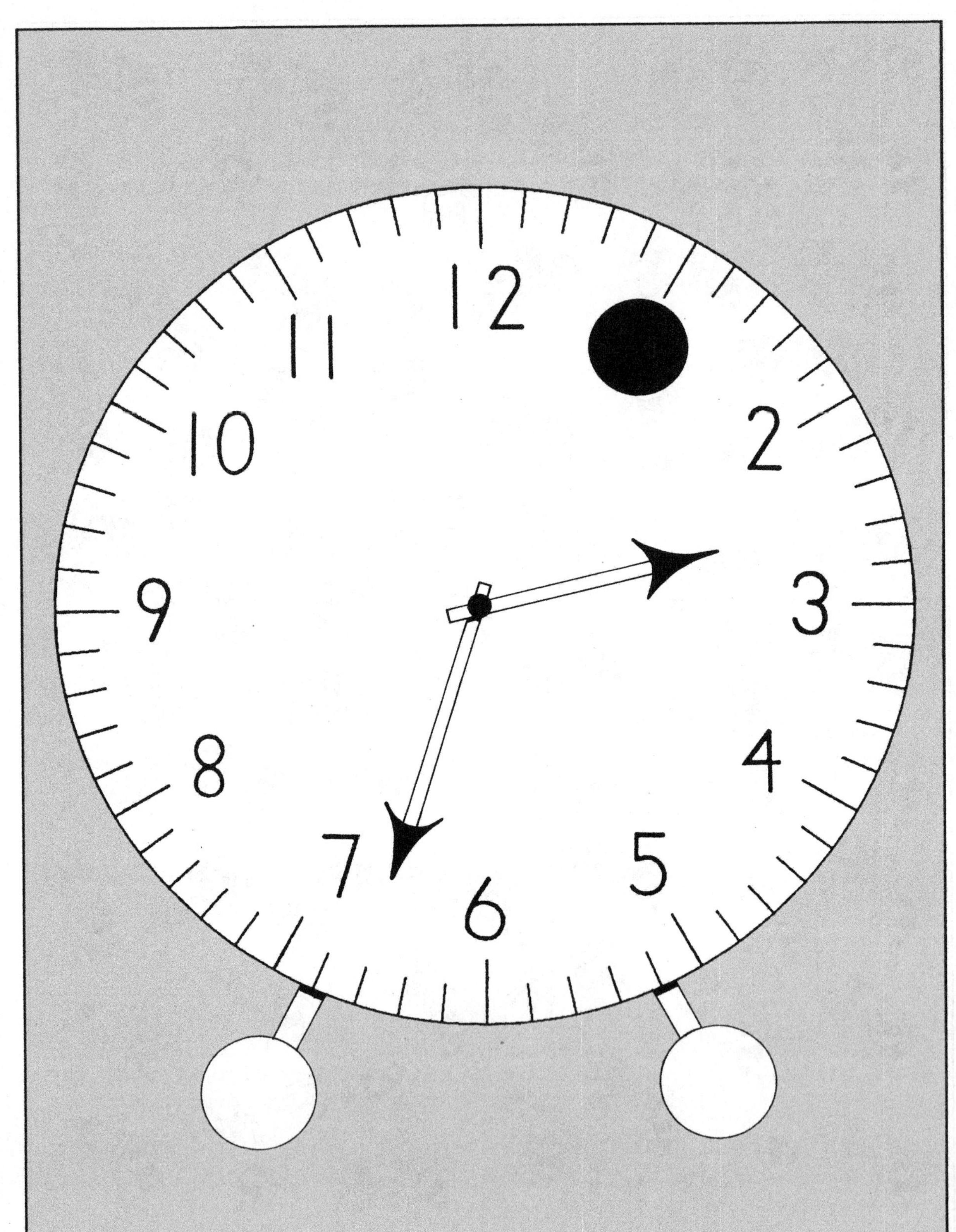
12
11
10
9
8
7
6
5
4
3
2

502	89	116	320
475	219	864	27
271	593	469	314
752	603	198	415
152	898	260	127
656	953	522	499

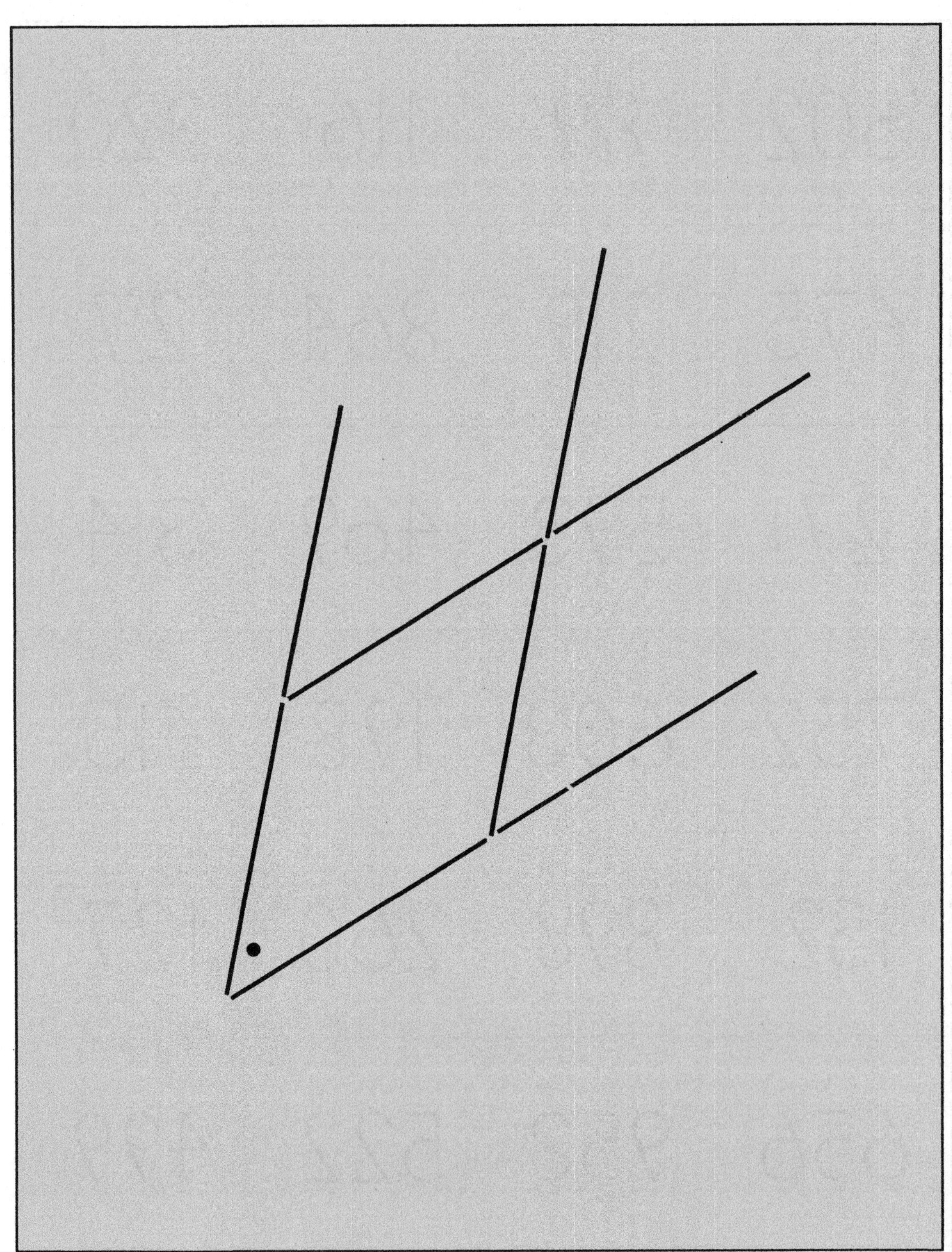

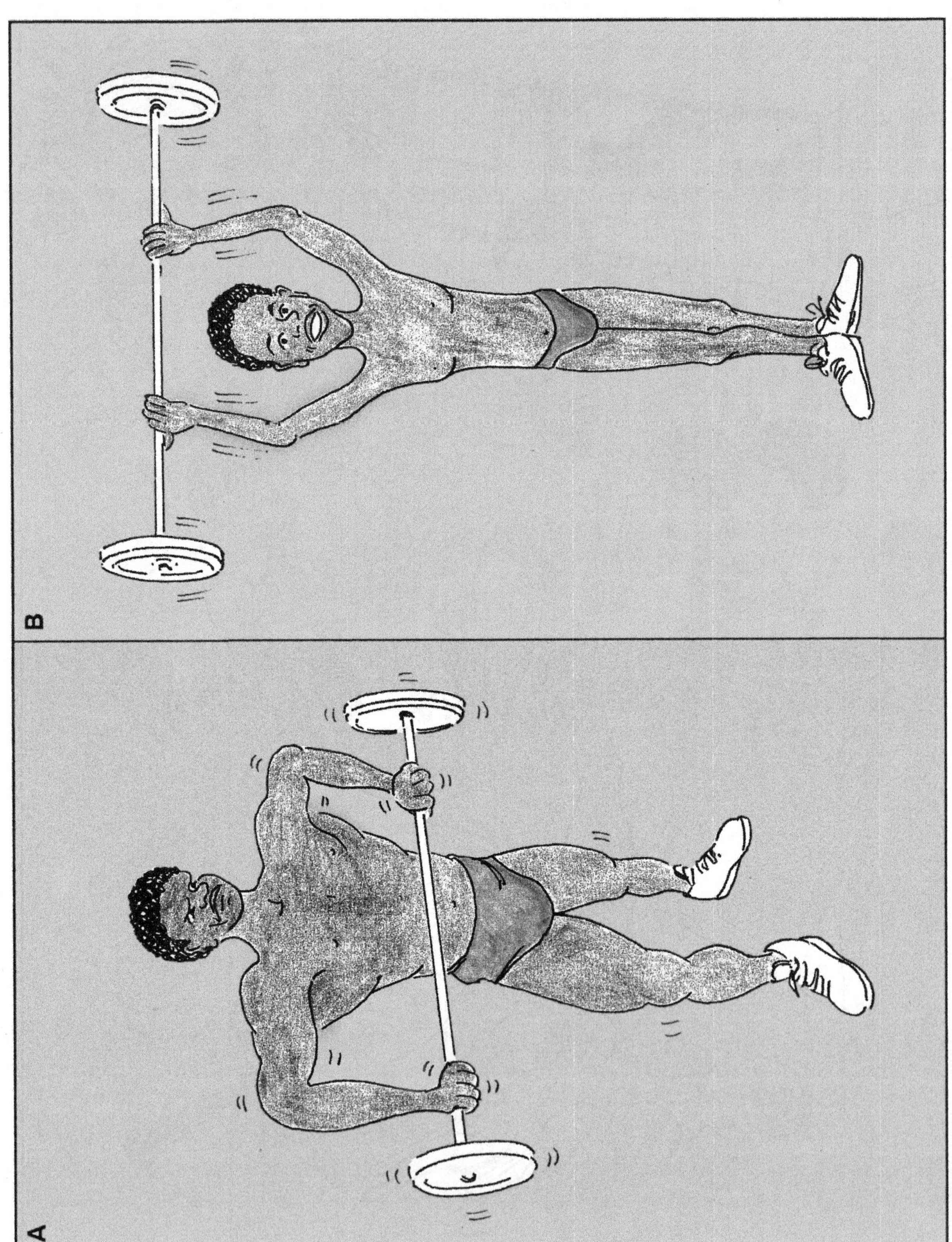
A
B

FINISH	36+16	60−9	30+50	56+14	70−11	89+6	34−5
75−14	25+30	45−13	71−8	50−25	19+31	72−13	56+6
49−20	16+11	31+17	16+16	55−16	74−15	29+31	38+20
75−25	89−23	67−20	76−38	32+45	46−18	5+29	47+15
28+26	24+62	47−15	64+16	55−30	27+33	82−10	74−18
38−10	16+21	19+31	15+16	61−17	19+19	72−19	57+10
80−21	26+30	91−30	49−21	18+39	63−19	14+67	52+15
15+15	17+18	19+23	80−16	61+17	18+44	54−19	40−18
71+18	59−25	51−19	60−35	16+39	87−15	36+36	67−10
26+36	80−11	72−11	76+14	83−13	69−9	42+15	START

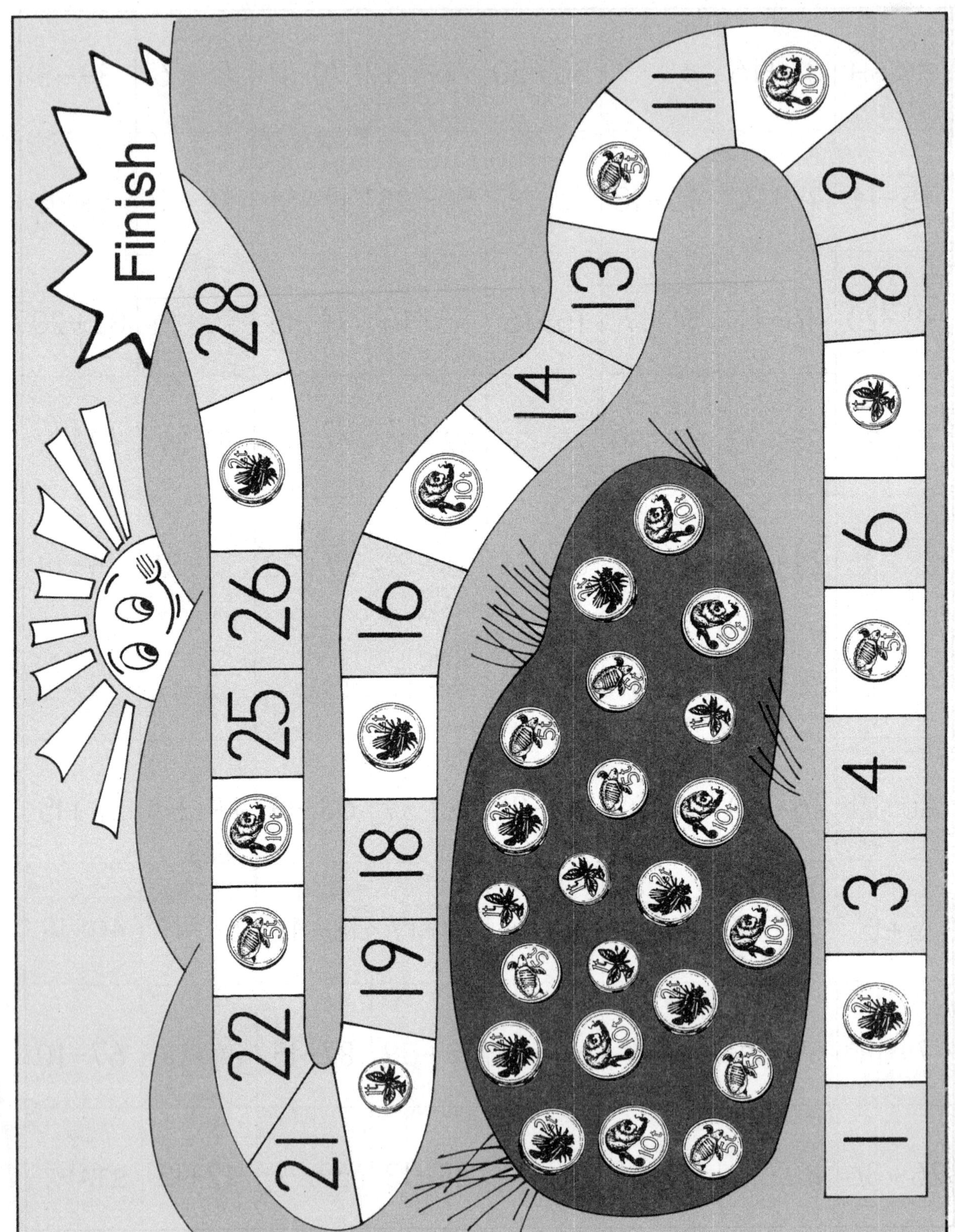
1
3
4
6
8
9
11
13
14
16
18
19
21
22
25
26
28
Finish